The Amazing Circle

Authors

Arthur J. Wiebe

Jim Wilson

Editors

Betty Cordel

Jim Wilson

Illustrator

Brenda Woods

Desktop Publishers

Tracey Lieder

Leticia Rivera

The ***AIMS Education Foundation*** *is a research and development organization dedicated to the improvement of the teaching and learning of mathematics and science through a meaningful integrated approach.*

This book contains materials developed by the AIMS Education Foundation. **AIMS** (**A**ctivities **I**ntegrating **M**athematics and **S**cience) began in 1981 with a grant from the National Science Foundation. The non-profit AIMS Education Foundation publishes hands-on instructional materials (books and the monthly magazine) that integrate curricular disciplines such as mathematics, science, language arts, and social studies. The Foundation sponsors a national program of professional development through which educators may gain both an understanding of the AIMS philosophy and expertise in teaching by integrated, hands-on methods.

ISBN **1-881431-72-X**

Printed in the United States of America

Table of Contents

Relating The Amazing Circle to the AIMS Model of Mathematics

The AIMS Model of Mathematics depicts the four environments in which mathematical activity takes place. The circle represents manipulative activities with real-world objects; the triangle consists of abstractions from and associations with real-world objects and events; the square depicts two-dimensional representations of three-dimensional objects and situations; and the hexagon represents those higher-order thinking skills by which the interrelationships among the others are discerned, explored, and utilized in problem solving.

Activities in *The Amazing Circle* normally begin in the square since the depicted circle and enclosed circular region are two-dimensional until transformed into a three-dimensional object by cutting it out. The Amazing Circle has a special definition since it consists of the circle, or boundary, and the region within.

Once cut out, the Amazing Circle becomes a three-dimensional object ready for manipulation. The circle in the Model is the arena where manipulation takes place. Students will fold and unfold the circle to form a broad range of interrelated shapes.

Activities represented by the triangular region include associating numerical and geometric properties with the real-world object, abstracting data from the real world, and naming. Here facts and numerical information are collected and learning of names and properties occurs. Observing, memorizing, recalling, identifying, and recording are predominant activities.

The higher-order thinking skills represented by the hexagon include identifying relationships, comparing, devising means for determining ratios, selecting and combining data to solve problems, hypothesizing and testing hypotheses.

The general movement will be from the square (cutting out the circle)
to the circle (folding and unfolding)
to the triangle (observing, etc.) and
to the hexagon (hypothesizing, problem solving).

A frequent final series of moves will be from the hexagon to the circle and back as hypotheses are formed, tested, and confirmed or revised.

A Model of Mathematics

Introducing The Amazing Circle

A plain circular sheet of paper contains a secret lode of rich geometric concepts and problem-solving opportunities few have explored. Pleasant surprises literally unfold at each new stage, serving as powerful motivators to extend the exploration. The secrets that lie hidden in this unassuming piece of paper can be extracted only by careful manipulation, observation, analysis, and calculation. The developing association between abstract concepts and their physical models enriches the entire learning experience. Connections among concepts are found in abundance.

Your students will be involved continuously as they fold, crease, unfold, observe, compare, identify, and draw conclusions regarding geometric properties and relationships. *The Amazing Circle* engages students in a hands-on, real-world, physical context that enhances understanding and retention of geometric concepts.

The Amazing Circle is the first of two monographs consisting of studies related to circles. In it students will find abundant opportunity to explore the nature and relationship of geometric figures, concepts, and terminology as embodied in a physical model. They will become thoroughly familiar with geometric terms, their physical representations, and the relationships through folding and unfolding circles.

The geometric content referenced in the teacher's guide indicates the large percentage of the standard curriculum covered in these investigations. Because all of the content referenced in *The Amazing Circle* is studied in context, the strong bond of interrelationships and connections that develops will serve students well by facilitating the retention of knowledge.

The processes and nature of the activities mirror the four environments contained in the AIMS Model of Mathematics. Hands-on folding and unfolding experiences (circle), naming, quantifying, and representing algebraically (triangle), illustrating (square) and formulating, generalizing, associating, reasoning, connecting, etc. (hexagon) are embedded in the content.

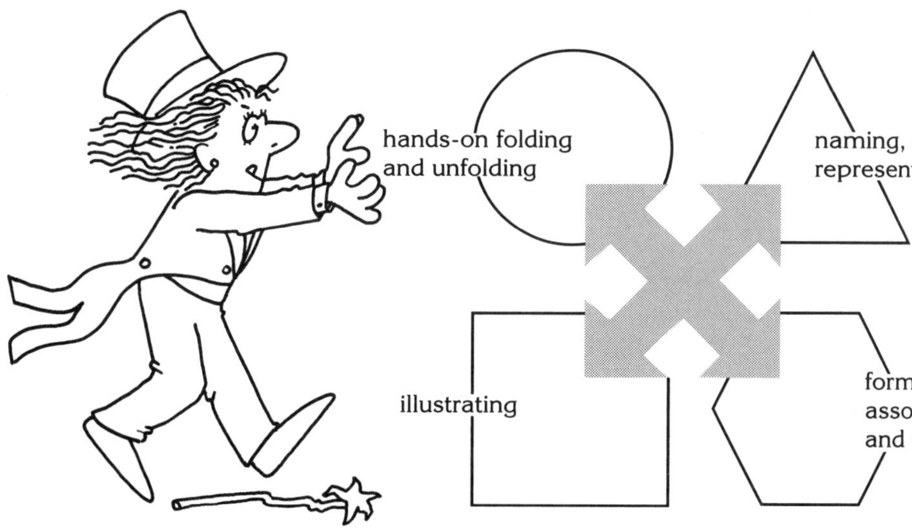

Students must be given adequate time to explore and form associations between the model on the one hand and the related terminology, concepts, connections, and relationships on the other. Students will benefit greatly from repeating the geometric explorations as often as necessary to ensure the development of firm bonding between the physical model and its associated meanings. Part of this repetition is built into and facilitated by the stages sequenced in the student pages.

For these activities make copies of the circle shown on page 3, using paper suitable for writing and mounting onto student pages. Each student and the teacher need a circle at each stage to manipulate, feel, fold, observe, compare, and discuss.

Precise terminology has been used where appropriate so that students are properly prepared for subsequent experiences where such precision will be expected. For example, a distinction is made between a *circle* and a *circular region*. A model of a circle is formed on paper by drawing a line that is a constant distance from a specific point called the center of the circle. A circular region consists of this circle and the area bounded by it.

The term "Amazing Circle" is given a special definition as consisting of the circle and the circular region enclosed by it.

Exploring the Geometric Concepts, Terminology, and Relationships of the Amazing Circle

Even though the sequence is presented here in its entirety, it is advisable to divide this exploration into parts to avoid introducing too many ideas at one time. It is important to provide sufficient time for thorough discussion and internalization of the concepts, terminology, and relationships that emerge at each stage. Instructions for folding are given in bold type. Sample questions (with answers shown in brackets) are provided for each stage. Most of these questions also appear on the student pages. Teachers are encouraged to discuss any or all of these questions and add their own.

Teachers may choose to introduce the Amazing Circle by just doing the step-by-step folding while keeping the use of unfamiliar terminology to a minimum. Such a preliminary activity helps students to become more adept at learning the folds. At the same time, the many surprises that are experienced along the way and the major surprise at the end will pique student interest and thereby create a positive mindset.

After students are comfortable with the folding and unfolding operations, they are ready to repeat the sequence, this time learning and making appropriate terminology associations. The questions under the heading *Identifications* are designed to enhance this process.

In the final sequence the questions under the heading *Relationships and Measurements* are studied. Students will be challenged to make careful comparisons and use higher-order reasoning skills to gain an understanding of the many relationships that exist within the circle.

The Amazing Circle is excellent for use with the whole class, small groups, or for individualized study. The formatting of successive student pages is designed to help students focus on a subset of concepts at a given time. The style used in making the completed student pages interactive will facilitate review.

The Amazing Circle

Duplicate as many circles as required.

Cut along the *outside* edge of the Amazing Circle so that the dark line of the circle is saved.

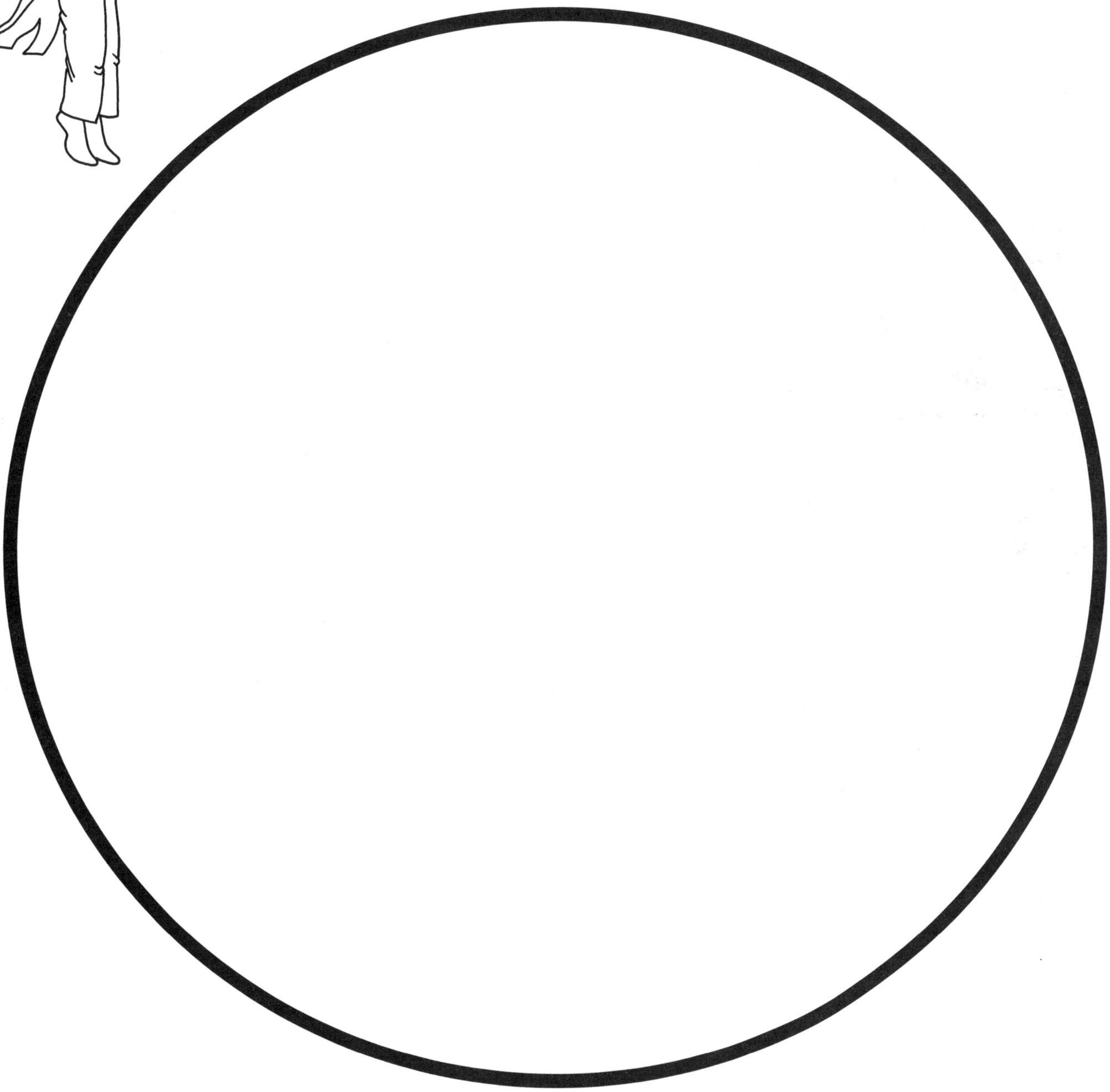

Exploring the *Amazing Circle* 1

Distribute the student page, *Amazing Circle Number 1.*

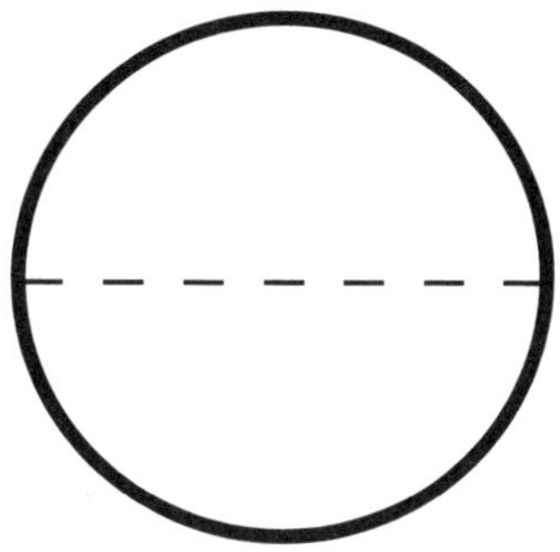

Fold the Amazing Circle in half and crease.

Identifications

a. What is this new shape called? [*Semi-circular region*]
b. What is its straight edge called? [*Diameter*]
c. In your own words, define a diameter. [the longest straight *line segment* with endpoints on the circle, a line segment through the center of the circle, etc.]
d. What is the curved edge of this new shape called? [*Semi-circle*]
e. How many semi-circles do you see? [Two]

Relationship

f. What part of the circular region lies within a semi-circular region? [One-half]

Exploration

g. Can you find a way to locate the center of the circle? [Fold a different semi-circular region and crease. The center of the circle is where the two creases intersect.]

Student Record

h. Have the students label their Amazing Circles as shown in the diagram. Instruct them to color or shade one of the semi-circular regions. Once completed, tell them to tape their Amazing Circle to their student page.

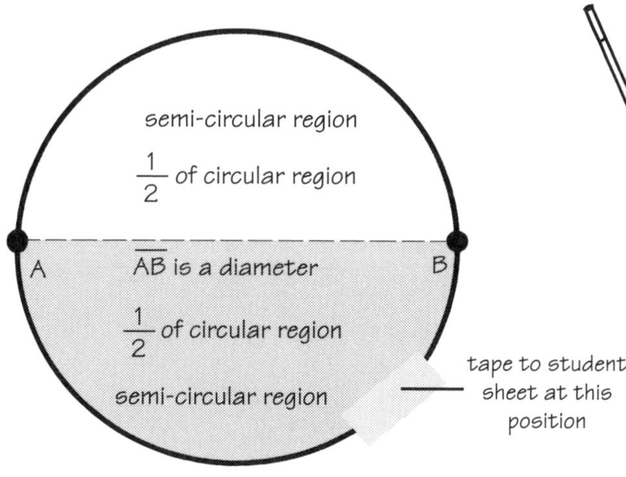

TEACHER GUIDELINES

Name Class Date

Amazing Circle Number 1

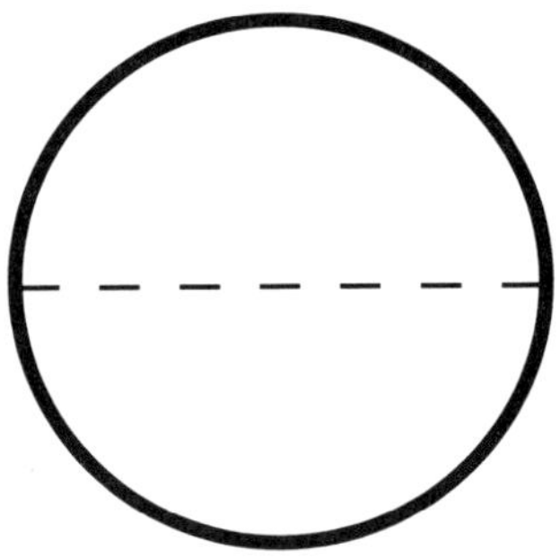

Cut out an Amazing Circle.
Fold in half and crease.

Identifications

a. What is this new shape called?

b. What is the straight edge called?

c. In your own words, define a diameter.

d. What is the curved edge of this new shape called?

e. How many semi-circles do you see?

Relationship

f. What part of the circular region lies within a semi-circular region?

Exploration

g. How can you locate the center of the circle?

tape here

h. Tape your Amazing Circle to the outlined circular region.

Exploring the *Amazing Circle* 2

Distribute the student page, *Amazing Circle Number 2.*

Fold the Amazing Circle in half (a). Open the circle and fold a second, different semi-circle. Open the circle. Mark the point where the diameters intersect.

Identifications

a. What can you say about the two diameters? [They intersect at the center of the circle, bisect each other, are congruent, create four radii, form four central angles, create four sectors, etc.]

b. What is the line segment equal to one-half a diameter called? [*Radius*]

c. The plural of radius is *radii.* How many radii do you see? [Four]

d. The diameters cut the circle into parts. What are they called? [*Arcs.* In this instance they are shorter than a semi-circle and are therefore called *minor* arcs. Those greater than a semi-circle are called *major* arcs.]

e. The two diameters created four regions. What are these called? [*Sectors*]

f. What makes up the boundary of a sector? [Two radii and an arc]

g. How many pairs of congruent sectors are there? [Two]

h. How many non-overlapping angles have been formed at the center? [Four. These are known as *central angles.*]

Relationships and Measurements

i. How do the shapes and sizes of opposite sectors compare? [They are *congruent*: same size, same shape.]

j. How does the measure of opposite angles compare? [The angles are congruent.]

k. What is the sum of any two adjacent central angles? [180° or a *straight angle*]

l. What shape do any two adjacent sectors form? [Semi-circular region]

m. What is the sum of any two adjacent arcs? [Semi-circle]

Student Record

n. Have the students label their Amazing Circles as shown in the diagram. Instruct them to color or shade one pair of opposite sectors. Once completed, have them tape their Amazing Circle to their page.

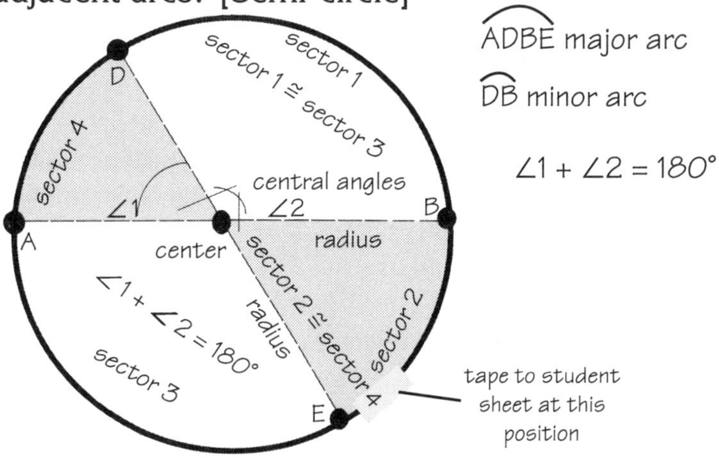

TEACHER GUIDELINES

Name Class Date

Amazing Circle Number 2

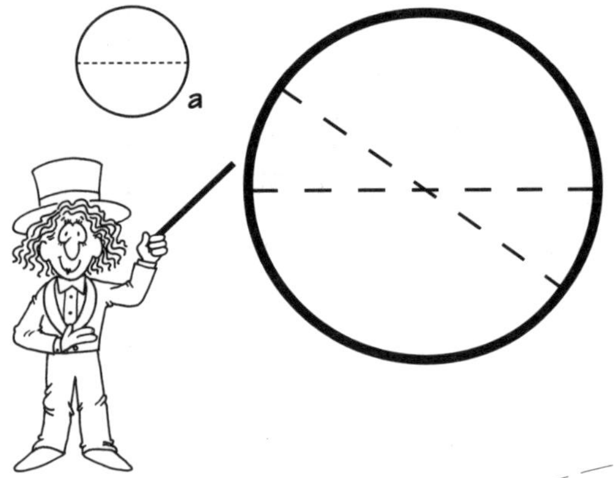

Cut out an Amazing Circle. Fold the circle in half (a). Open the circle and fold a second, different semi-circle. Open the circle. Mark the point where the diameters intersect.

Identifications

a. What can you say about the two diameters?

b. What is the line segment equal to one-half a diameter called?

c. The plural of radius is radii. How many radii do you see?

d. The diameters cut the circle into parts. What are those parts called?

e. The two diameters created four regions. What are these called?

f. What makes up the boundary of a sector?

g. How many pairs of of congruent sectors are there?

h. How many non-overlapping angles have been formed at the center?

Relationships and Measurements

i. How do the shapes and sizes of opposite sectors compare?

j. How does the measure of opposite angles compare?

k. What is the sum of any two adjacent central angles?

l. What shape do any two adjacent sectors form?

m. What is the sum of any two adjacent arcs?

n. Tape your Amazing Circle to the outlined circular region.

tape here

Exploring the *Amazing Circle* 3

Distribute the student page, *Amazing Circle Number 3.*

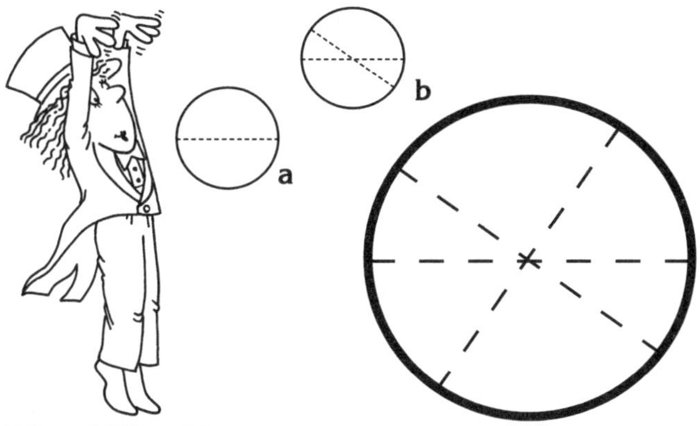

Fold an Amazing Circle in half (a). Fold a second (b) and then a third semi-circular region.

Identifications

a. What can you say about the three diameters? [They all pass through the center of the circle, are congruent, create six sectors, six radii, six minor arcs, six central angles, etc.]

b. What name is given to any line segment that passes through the center of the circle and has its endpoints on the circle. [Every such line segment must be a diameter.]

Relationships and Measurements

c. What shape do any three consecutive sectors form? [Semi-circular region]

d. What do any three consecutive arcs form? [Semi-circle]

e. What is the sum of any three consecutive central angles? [Straight angle, 180°]

f. How do opposite sectors compare? [They are congruent]

g. How many pairs of opposite angles are there? [Three if the required condition is that they do not overlap. Students might label them and then use combinations of two angles and their opposite two angles to further explore this question.]

h. How does the angle measure of opposite angles compare? [They are congruent.]

Student Record

i. Have the students label their Amazing Circles as shown in the diagram. Instruct them to color or shade opposite sectors. Once completed, have them tape their Amazing Circle to their page.

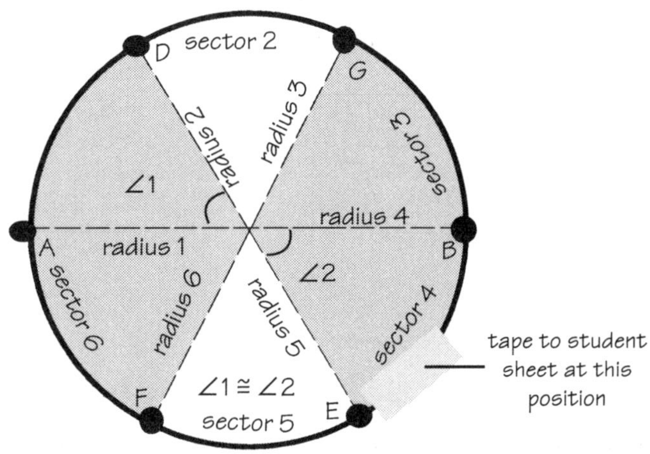

TEACHER GUIDELINES

Name Class Date

Amazing Circle Number 3

Cut out an Amazing Circle. Fold the circle in half (a). Open the circle and fold a second (b) and then a third different semi-circular region.

Identifications

a. What can you say about the three diameters?

b. What name is given to any line segment that passes through the center of the circle and has its endpoints on the circle?

Relationships and Measurements

c. What shape do any three consecutive sectors form?

d. What do any three consecutive arcs form?

e. What is the sum of any three consecutive central angles?

f. How do opposite sectors compare?

g. How many pairs of opposite angles are there?

h. How does the angle measure of opposite angles compare?

tape here

i. Tape your Amazing Circle to the outlined circular region.

Exploring the *Amazing Circle* 4

Distribute the student page, *Amazing Circle Number 4.*

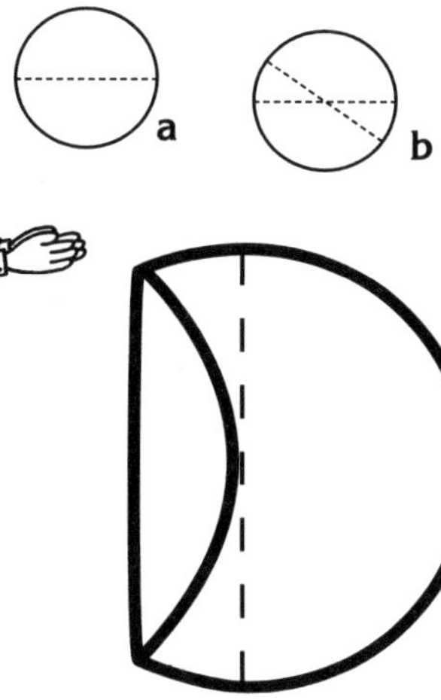

Fold the Amazing Circle in half and crease (a). Fold a second and different diameter to locate the center of the circle. Open the circle (b). Bring one point on the circle to the center and fold, creating a sharp crease.

Identifications

a. What is the straight line segment formed by the fold called? [A *chord.* All creases with endpoints on the circle are called chords of the circle. A diameter is a chord that passes through the center of the circle.]

b. What constitutes the boundary of this new figure? [A chord and a major arc.]

c. What constitutes the boundary of the part that has been folded? [A chord and a minor arc.]

Student Record

d. Have the students label their Amazing Circles as shown in the diagram. Instruct them to color or shade the shape bounded by the chord and the major arc. Once completed, have them tape their Amazing Circle to their page.

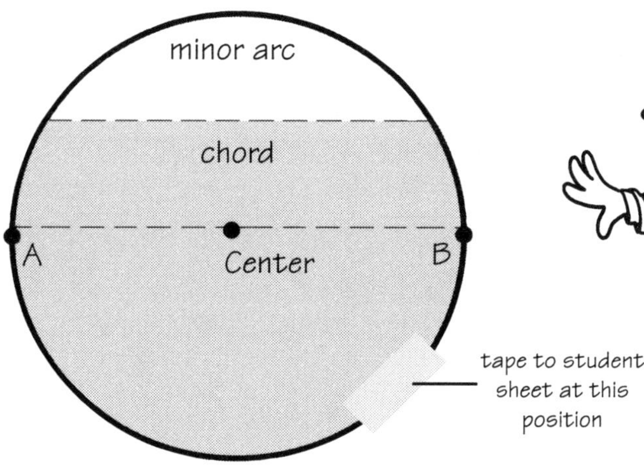

TEACHER GUIDELINES

Name Class Date

Amazing Circle Number 4

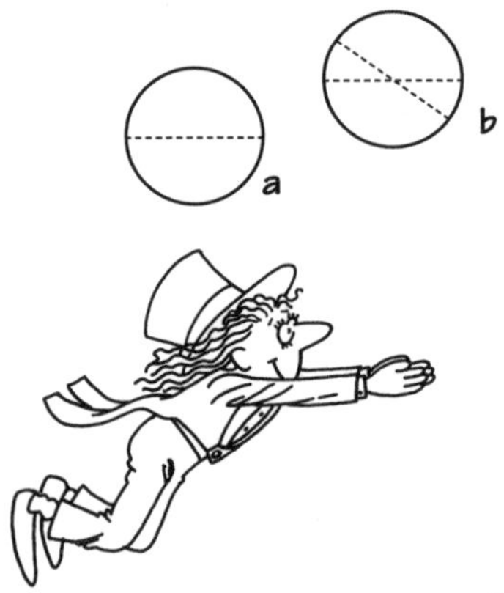

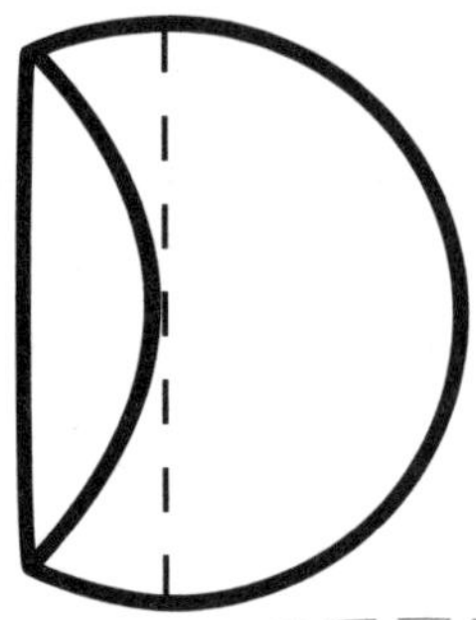

Cut out an Amazing Circle. Fold two diameters (a and b) to locate the center of the circle. Bring one point on the circle to the center and fold, creating a sharp crease.

Identifications

a. What is the straight line segment formed by the fold called?

b. What constitutes the boundary of this new figure?

c. What constitutes the boundary of the folded over part?

tape here

d. Tape your Amazing Circle to the outlined circular region.

Exploring the *Amazing Circle* 5

TEACHER GUIDELINES

Distribute the student page, *Amazing Circle Number 5.*

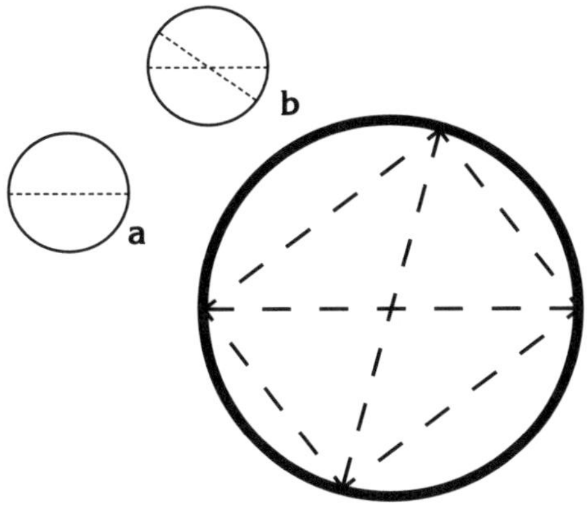

Fold the Amazing Circle in half and crease (a). Open the Amazing Circle and fold a second and different semicircle that divides the Amazing Circle in divisions other than fourths (b). Mark the point where the diameters intersect. Fold the chords connecting adjacent ends of the diameters.

Identifications

a. What is the shape that is formed by the chords called? [*Rectangle*]
b. What is the name given to each diameter? [*Diagonal*]

Relationships and Measurements

c. How do the opposite sides of the rectangle compare? [They are congruent.]
d. How do opposite angles formed by the diagonals of the rectangle compare? [They are congruent.]
e. Describe how the diagonals of the rectangle intersect. [The diagonals bisect each other.]
f. What shapes are formed by the diagonals in the interior of the rectangle? [*Isosceles* and *right triangles.*]
g. How many triangles do you see in the rectangle? [Eight]

h. How many *pairs* of congruent triangles are formed and what types of triangles are they? [Four, two isosceles and two right triangles]
i. Of the isosceles triangles, how many have only acute angles? [Two] How many of the isosceles triangles have one obtuse angle? [Two]
j. Are the sides of the rectangle opposite the obtuse angles the longer or shorter sides of the rectangle? [Longer]
k. How do the four angles of the rectangle with vertices on the circle compare? [They are congruent right angles.]

TEACHER GUIDELINES

Student Record

l. Have the students label their Amazing Circles as shown in the diagram. Instruct them to color or shade in the rectangle. Have them also write on their circle the congruent line segments, angles, and triangles. Once completed, tell them to tape their Amazing Circle to their page.

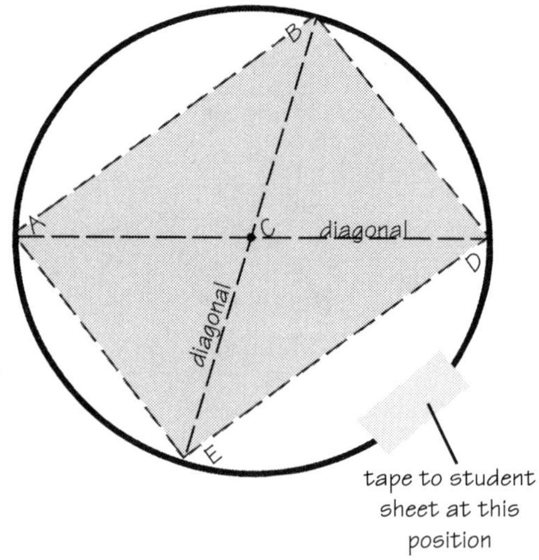

congruent line segments	congruent triangles	congruent angles
$\overline{AC} \cong \overline{DC}$	$\Delta ABC \cong \Delta EDC$	$\angle ABD \cong \angle BDE$ = right angle
$\overline{AE} \cong \overline{BD}$	$\Delta ACE \cong \Delta BCD$	$\angle DEA \cong \angle EAB$ = right angle
$\overline{AB} \cong \overline{ED}$	$\Delta ABD \cong \Delta AED$	$\angle ACB \cong \angle DCE$
$\overline{BC} \cong \overline{EC}$	$\Delta ABE \cong \Delta ADE$	$\angle ACE \cong \angle BCD$
$\overline{AD} \cong \overline{BE}$		

Exploration

m. **Start with a new Amazing Circle. Fold the Amazing Circle so that the rectangle formed by two intersecting diameters is a square. Label any relationships and measurements you discover on your Amazing Square. [Fold the Amazing Circle in half and then in half again. This produces four congruent sectors. Fold the chords connecting adjacent ends of the diameters.]**

The Exploration Student Record

n.

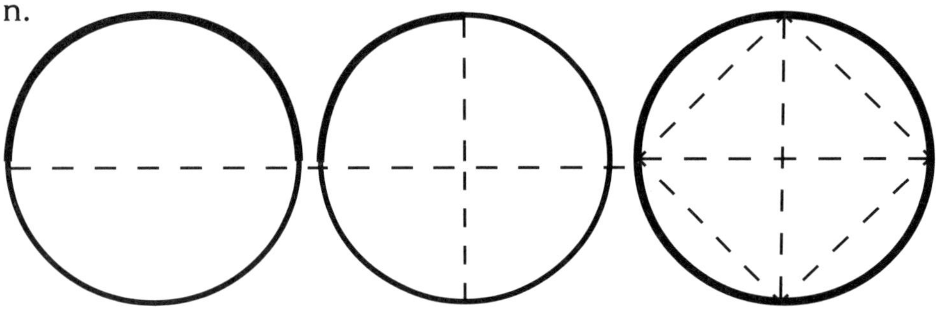

the first fold the second fold the circle unfolded

congruent line segments
$\overline{AD} \cong \overline{BE}$
$\overline{AD} \perp \overline{BE}$ and C is the midpoint of $\overline{AD}$ and $\overline{BE}$
$\overline{AC} \cong \overline{DC}$
$\overline{BC} \cong \overline{EC}$
$\overline{AB} \cong \overline{BD} \cong \overline{DE} \cong \overline{EA}$

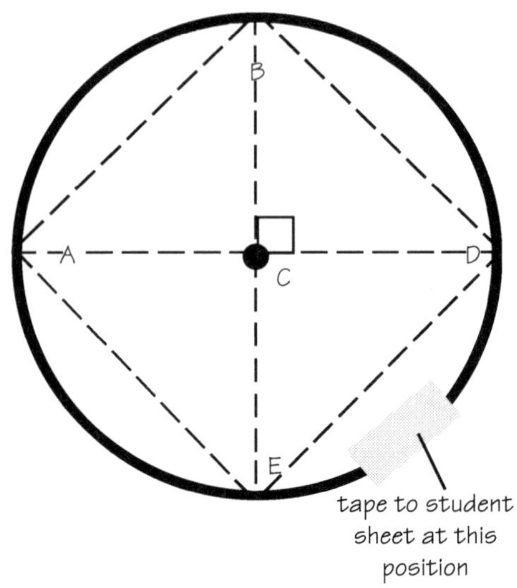

congruent angles
$\angle ABD \cong \angle BDE \cong \angle DEA \cong \angle EAB \cong$ right angle
$\angle ACB \cong \angle BCD \cong \angle DCE \cong \angle ECA \cong$ right angle
$\angle CBD \cong \angle CDB \cong \angle CDE \cong \angle CED \cong \angle CEA \cong \angle CAE \cong \angle CAB \cong \angle CBA$

congruent triangles
$\Delta ACB \cong \Delta BCD \cong \Delta DCE \cong \Delta ECA$
$\Delta ABE \cong \Delta ADE \cong \Delta ABD \cong \Delta BED$
[all are isosceles right triangles]

Name Class Date

Amazing Circle Number 5

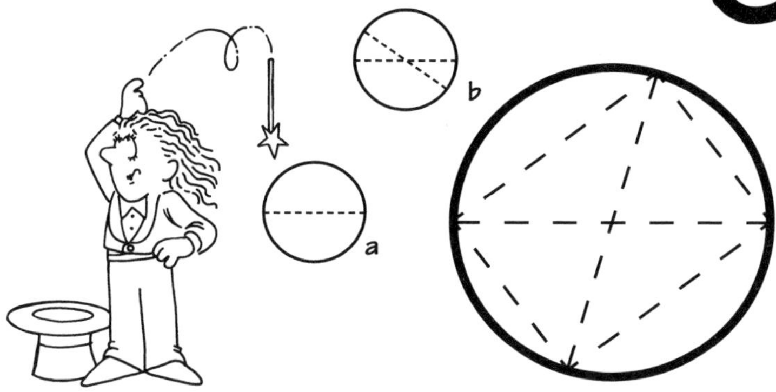

Fold the Amazing Circle in half and crease (a). Open the Amazing Circle and fold a second and different semi-circle that divides the Amazing Circle in divisions other than fourths (b). Mark the point where the diameters intersect. Fold the chords connecting adjacent ends of the diameters.

Identifications

a. What is the shape that is formed by the chords called?

b. What is the name given to each diameter?

Relationships and Measurements

c. How do the opposite sides of the rectangle compare?

d. How do opposite angles formed by the diagonals of the rectangle compare?

e. Describe how the diagonals of the rectangle intersect.

f. What shapes are formed by the diagonals in the interior of the rectangle?

g. How many triangles do you see in the rectangle?

h. How many pairs of congruent triangles are formed and what types of triangles are they?

i. How many of the isosceles triangles have only acute angles?

How many of the isosceles triangles have one obtuse angle?

j. Are the sides of the rectangle opposite obtuse angles the longer or shorter sides of the rectangle?

k. How do the four angles of the rectangle with vertices on the circle compare?

l. Tape your Amazing Circle to the outlined circular region.

tape one here

tape one here

Exploration

m. Start with a new Amazing Circle. Fold the Amazing Circle so that the rectangle formed by two intersecting diameters is a square. Label any relationships and measurements you discover on your Amazing Circle.

n. Tape your Amazing Circle to the outlined circular region.

Exploring the *Amazing Circle* 6

Distribute the student page, *Amazing Circle Number 6.*

Fold two different diameters to locate the center of the Amazing Circle (a). Bring one point on the circle to the center and fold, making a sharp crease (b). Make a similar fold so that one end of the new chord meets an end of the first chord and crease. (The figure should resemble an ice cream cone.)

Identifications

a. What elements form the boundary of this figure? [Two chords and a minor arc.]

b. What is the angle formed by the two chords called? (*Inscribed angle.* The chords forming an inscribed angle have their end points on the circle.]

c. What is the minor arc called? [the *intercepted arc*]

Relationships and Measurements

d. What part of the circle is in the minor arc? [One-third]

e. What is the measure of this arc if the measure of the circle is360 degrees? [120 degrees]

f. What is the measure of the inscribed angle? [60 degrees. (Students could use a protractor to find the answer.)]

g. How does the measure of the inscribed angle compare to that of the intercepted arc? [It is one-half that of the intercepted arc. The measures are 60 degrees and 120 degrees, respectively.]

Student Record

h. Have the students label their Amazing Circle as shown in the diagram. Instruct the students to color or shade the region bounded by the two chords and the minor arc. Once completed, have them tape their Amazing Circle to their page.

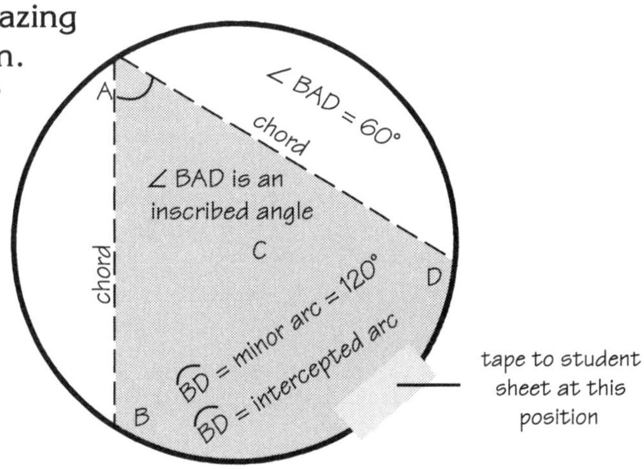

TEACHER GUIDELINES

Name Class Date

Amazing Circle Number 6

Cut out an Amazing Circle. Fold two different diameters to locate the center of the circle (a). Bring one point on the circle to the center and fold, creating a sharp crease (b). Make a similar second fold so that one end of the new chord meets an end of the first chord and crease. (The figure should resemble an ice cream cone.)

Identifications

a. What elements form the boundary of this figure?

b. What is the angle formed by the two chords called?

c. What is the minor arc called ?

Relationships and Measurements

d. What part of the circle is in the minor arc?

e. If the measure of the circle is 360 degrees, what is the measure of this arc ?

f. What is the measure of this inscribed angle?

g. How does the measure of the inscribed angle compare to that of the intercepted arc?

tape here

h. Tape your labeled and colored Amazing Circle to the outlined circular region.

Exploring the *Amazing Circle* 7

Distribute the student page, *Amazing Circle Number 7.*

Fold two different diameters to locate the center of the Amazing Circle (a). Bring one point on the circle to the center and fold, making a sharp crease (b). Make a similar fold so that one end of the new chord meets an end of the first chord and crease (c). Make a third fold to the center so that the ends of the new chord meet the remaining ends of the previous two chords. Crease.

Identifications

a. What is the shape that is formed called? [*Equilateral triangle*]

b. What is the point where the ends of two chords meet called? [A *vertex*. Plural: *vertices*]

Relationships and Measurements

c. How do the measures of the sides of the equilateral triangle compare? [They are congruent.]

d. How do the angles of the equilateral triangle compare? [They are congruent.]

e. Is the area of the equilateral triangle more or less than one-half the area of the circular region? [Less than one-half since the triangle is completely covered by the folded over portion with some overlapping areas.]

f. If the measures of the three angles of any triangle are added, the sum is 180°. What is the measure of each of these angles? [60 degrees.]

g. How many vertices do you see? [Three]

Student Record

h. Have the students label their Amazing Circles as shown in the diagram. Instruct the students to color in or shade the equilateral triangle. Once completed, have them roll-tape their Amazing Circle to their page.

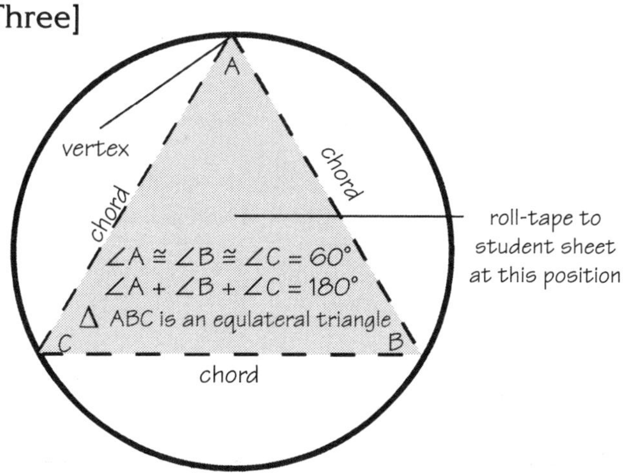

TEACHER GUIDELINES

Name Class Date

Amazing Circle Number 7

Fold two different diameters to locate the center of the Amazing Circle (a). Bring one point on the circle to the center and fold, making a sharp crease (b). Make a similar fold so that one end of the new chord meets an end of the first chord and crease (c). Make a third fold to the center so that the ends of the new chord meet the remaining ends of the previous two chords. Crease.

Identifications

a. What is the shape that is formed called?

b. What is the point where the ends of two chords meet called?

Relationships and Measurements

c. How do the measures of the sides of the equilateral triangle compare?

d. How do the angles of the equilateral triangle compare?

e. Is the area of the equilateral triangle more or less than one-half the area of the circular region?

roll-tape here

f. If the measures of the three angles of any triangle are added, the sum is 180°. What is the measure of each of these angles?

g. How many vertices do you see?

h. Tape your labeled and colored-in Amazing Circle to the outlined circular region.

Exploring the *Amazing Circle* 8

Distribute the student page, *Amazing Circle Number 8.*

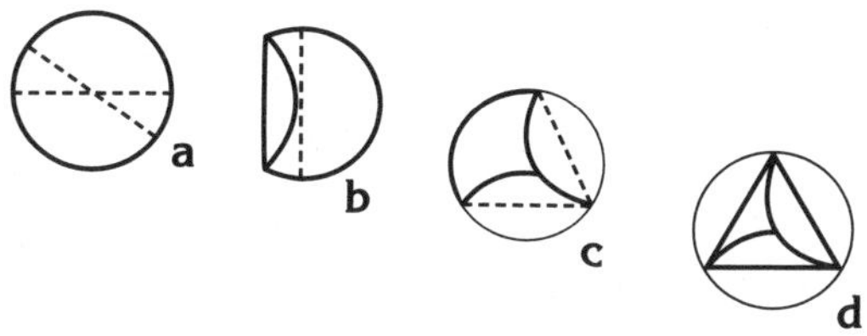

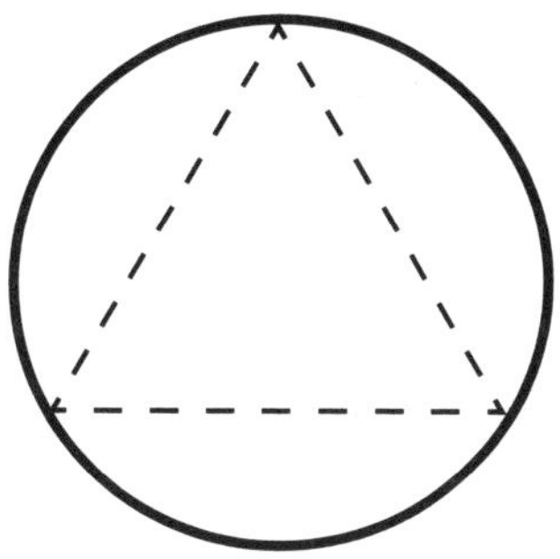

Fold the Amazing Circle to the large equilateral triangle (a, b, c, d). Open the circle. Note that the equilateral triangle is "inscribed" in the circle; that is, each vertex is on the circle.

Relationship

a. How does the area of the triangle compare with the area of the circle outside the triangle? [The area of the triangle is less since the outside will cover all of the triangle with some overlapping parts.]

Student Record

b. Have the students label their Amazing Circles as shown in the diagram. Instruct the students to color in or shade the equilateral triangle and the three pieces of the circle outside of the triangle. Once completed, have them tape their Amazing Circle to their page.

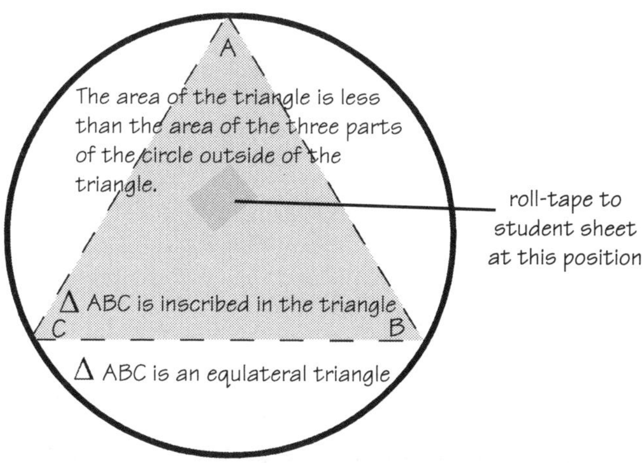

TEACHER GUIDELINES

Name Class Date

Amazing Circle Number 8

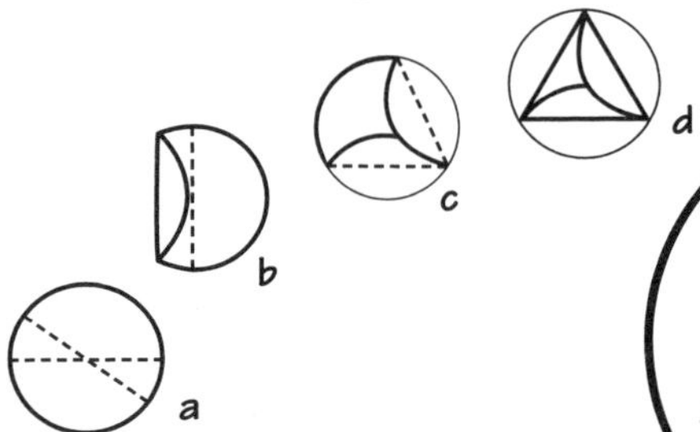

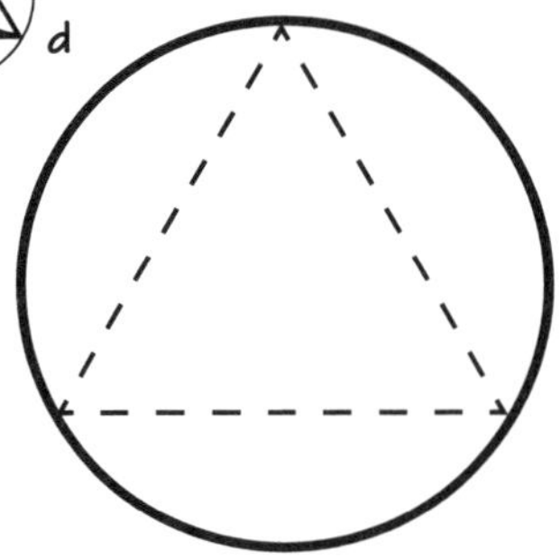

Fold the Amazing Circle to the large equilateral triangle (a, b, c, d). Open the Amazing Circle. Note that the equilateral triangle is "inscribed" in the circle; that is, each vertex is on the circle.

Relationship

a. How does the area of the triangle compare with the area of the circle outside the triangle?

roll-tape here

b. Tape your labeled and colored-in paper Amazing Circle to the outlined circular region.

Exploring the *Amazing Circle* 9

Distribute the student page, *Amazing Circle Number 9.*

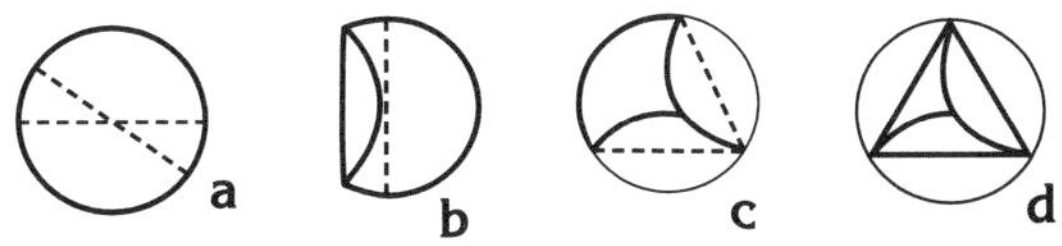

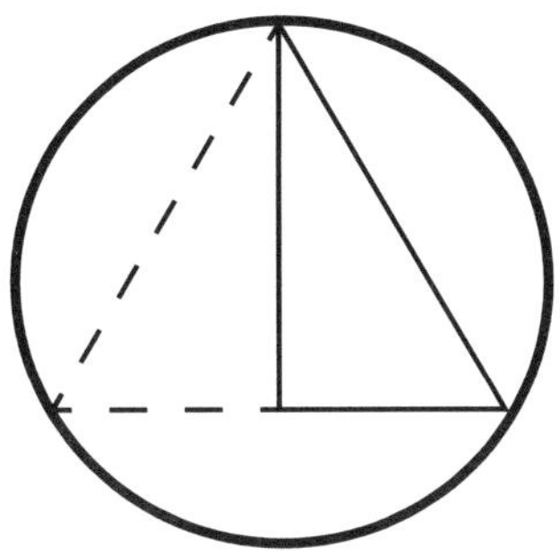

Fold the Amazing Circle to the equilateral triangle (a, b, c, d). Find the midpoint of one side of the triangle by folding. Make a fold passing through this midpoint and the opposite vertex.

Identifications

a. What type of triangle is formed? [*Right triangle*]
b. What is the longest side of a right triangle called? [*Hypotenuse*]
c. What are the two shorter sides called? [*Legs*]

Relationships and Measurements

d. How does the length measure of the shorter leg compare to the measure of the length of the hypotenuse? [One-half as long]
e. What is the measure of the angle opposite the hypotenuse? [90°]
f. What is the measure of the angle opposite the shorter leg? [30°]
g. What is the measure of the angle opposite the longer leg? [60°]
h. How does the area of the right triangle compare with that of the equilateral triangle? [It is one-half that of the equilateral triangle.]
i. How does the folded-over triangle compare to the triangle under it? [They are congruent right triangles.]

Student Record

j. Have the students label their Amazing Circles as shown in the diagram. Instruct the students to color in or shade the right triangle. Once completed, have them roll-tape their Amazing Circle to their page.

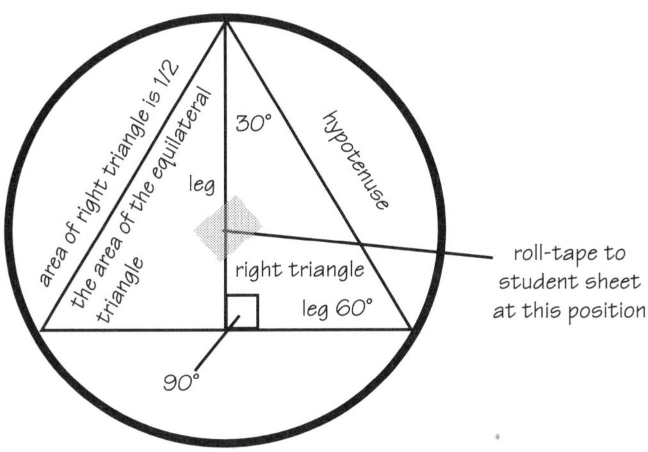

TEACHER GUIDELINES

Name Class Date

Amazing Circle Number 9

Fold the Amazing Circle to the equilateral triangle (a, b, c, d). Find the midpoint of one side of the triangle by folding. Make a fold passing through this midpoint and the opposite vertex.

Identifications

a. What type of triangle is formed?

b. What is the longest side of a right triangle called?

c. What are the two shorter sides called?

Relationships and Measurements

d. How does the length measure of the shorter leg compare to the length measure of the hypotenuse?

e. What is the measure of the angle opposite the hypotenuse?

roll-tape here

f. What is the measure of the angle opposite the shorter leg?

g. What is the measure of the angle opposite the longer leg?

h. How does the area of the right triangle compare with that of the equilateral triangle?

i. How does the folded-over triangle compare to the triangle under it?

j. Tape your labeled and colored-in Amazing Circle to the outlined circular region.

Exploring the *Amazing Circle* 10

Distribute the student page, *Amazing Circle Number 10.*

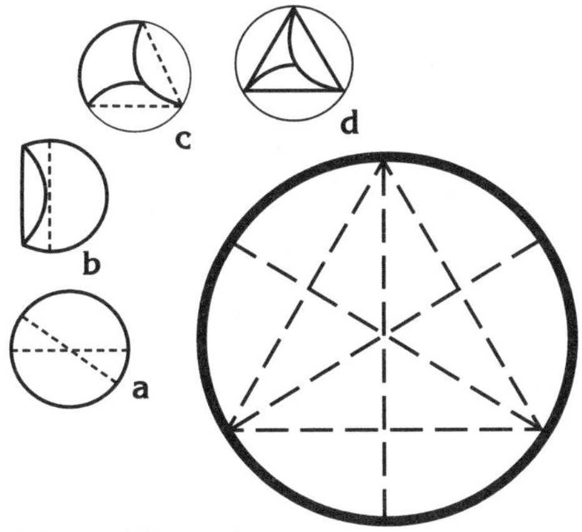

Fold the Amazing Circle to the equilateral triangle (a, b, c, d). Find the mid-points of the three sides. Make a fold through these mid-points and the opposite vertices. Open to the circle. Highlight the three diameters with endpoints on the vertices and passing through the respective mid-points of the sides of the triangle.

Identifications

a. What names can be given to the line segments connecting mid-points and their opposite vertices? [*Medians, perpendicular bisectors of the sides, bisectors of the angles at the vertices*]

b. What is a median? [A line segment connecting a vertex with the mid-point of the opposite side of a triangle]

c. Where do the medians in the triangle intersect? [Center of the circle]

Relationships and Measurements

d. What are the length ratios of each of the following line segments:
   (1) vertex-intersection to intersection-midpoint? [2:1]
   (2) vertex-intersection to diameter? [1:2]
   (3) intersection-midpoint to diameter? [1:4]
   (4) intersection-midpoint to median? [1:3]
   (5) vertex-intersection to median? [2:3]

Exploration

e. Construct various triangles that are not equilateral. Construct the medians. What is the length ratio of the vertex-intersection line segment to the median? [2:3]

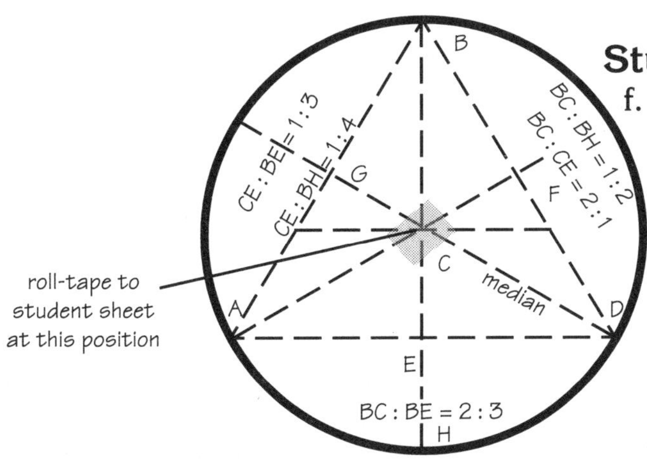

Student Record

f. Have the students label their Amazing Circles as shown in the diagram. Instruct the students to color in or shade the equilateral triangle. Tell them to label each median, the center of the circle, and each mid-point. They can also record each ratio. Once completed, have them roll-tape the center of their Amazing Circle to their page.

TEACHER GUIDELINES

Name Class Date

Amazing Circle Number 10

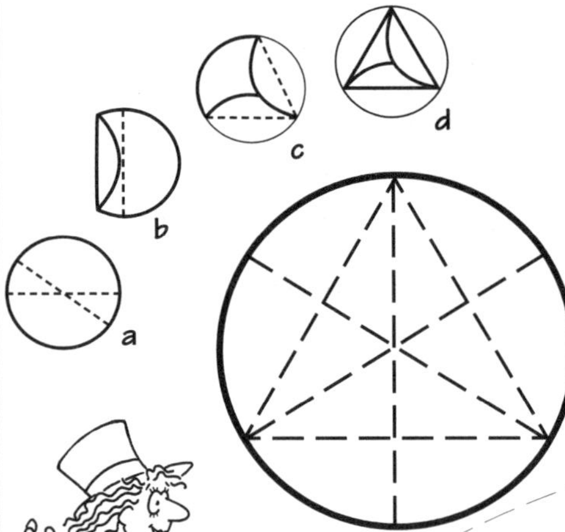

Cut out and fold the Amazing Circle to the equilateral triangle (a, b, c, d). Find the mid-points of the three sides. Make a fold through these mid-points and the opposite vertices. Open to the circle. Highlight the three diameters with endpoints on the vertices and passing through the respective mid-points of the sides of the triangle.

Identifications

a. What names can be given to the line segments connecting mid-points and their opposite vertices?

b. What is a median?

c. Where do the medians in the triangle intersect?

roll-tape here

Relationships and Measurements

d. What are the length ratios of each of the following line segments:

(1) vertex-intersection to intersection-midpoint?

(2) vertex-intersection to diameter?

(3) intersection-midpoint to diameter?

(4) intersection-midpoint to median?

(5) vertex-intersection to median?

Exploration

e. Construct various triangles that are not equilateral. Construct the medians. What is the length ratio of the vertex-intersection line segment to the median?

f. Tape your labeled and colored-in Amazing Circle to the outlined circular region.

Exploring the *Amazing Circle* 11

Distribute the student page, *Amazing Circle Number 11.*

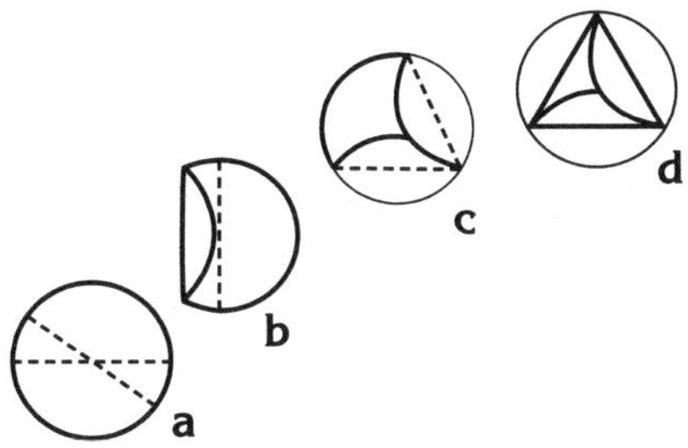

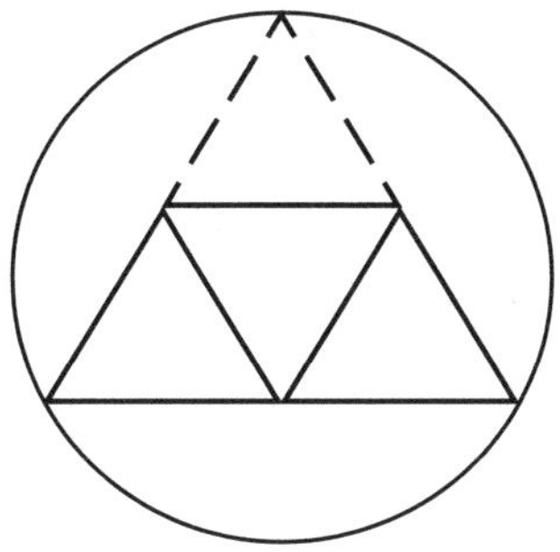

Fold an Amazing Circle to the equilateral triangle (a, b, c, d). Bring one vertex of the equilateral triangle to the midpoint of the opposite side and crease.

Identifications

a. What type of figure is formed? [*Isosceles trapezoid*]
b. How many triangles are visible? [Three]
c. What type of triangles are they? [Equilateral]

Relationships and Measurements

d. How do the lengths of the two non-parallel sides of the trapezoid compare? [They are congruent.]
e. How do lengths of the two parallel sides compare? [One is twice as long as the other.]
f. How do these triangles compare? [They are congruent.]
g. How does the area of the trapezoid compare with that of the original equilateral triangle? Explain your reasoning. [It is three-fourths that of the original triangle because three small triangles form the trapezoid compared to four for the large triangle.]
h. How does the area of the large equilateral triangle compare with that of the trapezoid? [It is 4/3 as large, the inverse relationship.]
i. How does the area of one of the small equilateral triangles compare with that of the trapezoid? [It is 1/3 the area of the trapezoid.]
j. How does the area of one of the small triangles compare with that of the original equilateral triangle? [It has 1/4 the area.]
k. How does the perimeter of the trapezoid compare with that of the large equilateral triangle? [It is 5/6 that of the large triangle.]
l. How does the perimeter of the trapezoid compare with that of one of the small equilateral triangles? [It is 5/3 as long.]
m. How does the perimeter of the small triangle compare with that of the large triangle? [It is 1/2 as long.]
n. What are the measures of the angles of the trapezoid? [Two are 60˚ and two are 120˚]
o. What is the sum of the four angles of the trapezoid? [360˚]

Student Record

p. Have the students label their Amazing Circles as shown in the diagram. Instruct the students to color in or shade the isosceles trapezoid. Once completed, have them roll-tape the center of their Amazing Circle to their page. They will then be able to fold to the trapezoid and unfold back to the equilateral triangle.

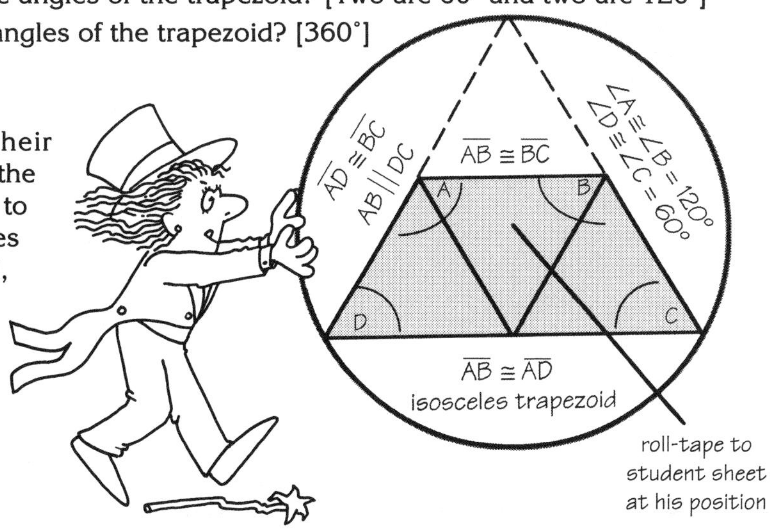

TEACHER GUIDELINES

Name Class Date

Amazing Circle Number 11

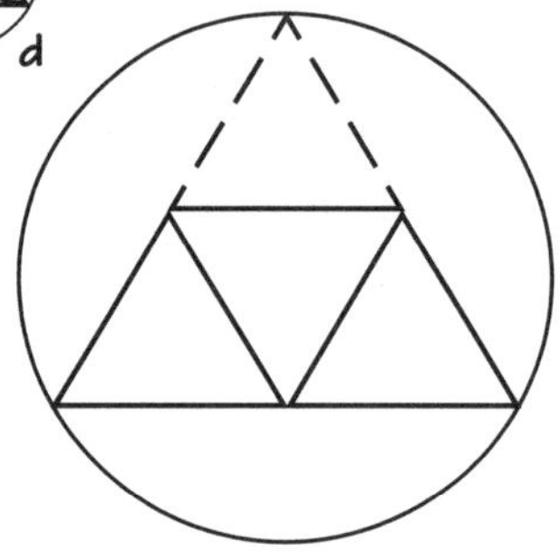

Cut out an Amazing Circle. Fold to an equilateral triangle (a, b, c, d). Bring one vertex of the equilateral triangle to the midpoint of the opposite side and crease.

Identifications

a. What type of figure is formed?

b. How many triangles are visible?

c. What type of triangles are they?

Relationships and Measurements

d. How do the lengths of the two non-parallel sides of the trapezoid compare?

e. How do lengths of the two parallel sides compare?

f. How do these triangles compare?

g. How does the area of the trapezoid compare with that of the original equilateral triangle? Explain your reasoning.

h. How does the area of the large equilateral triangle compare with that of the trapezoid?

roll-tape here

i. How does the area of one of the small equilateral triangles compare with that of the trapezoid?

j. How does the area of one of the small triangles compare with that of the original equilateral triangle?

k. How does the perimeter of the trapezoid compare with that of the large equilateral triangle?

l. How does the perimeter of the trapezoid compare with that of one of the small equilateral triangles?

m. How does the perimeter of the small triangle compare with that of the large triangle?

n. What are the measures of the angles of the trapezoid?

o. What is the sum of the four angles of the trapezoid?

p. Tape your labeled and colored-in Amazing Circle to the outlined circular region.

TEACHER GUIDELINES

Exploring the *Amazing Circle* 12

Distribute the student page, *Amazing Circle Number 12.*

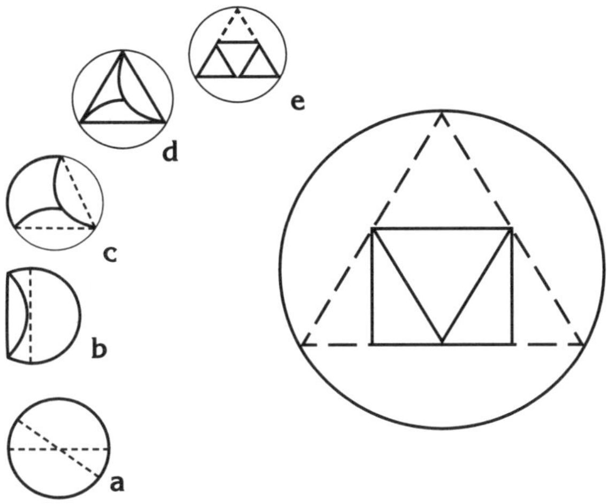

Fold an Amazing Circle to the isosceles trapezoid (a, b, c, d, e). Fold the second and third vertices of the large triangle to the same mid-point.

Identifications

a. What type of figure is formed? [Rectangle]

Relationships and Measurements

b. What can you say about the two pairs of opposite sides? [They are congruent and parallel.]

c. What are the ratios of the following:
   (1) height of rectangle to height of original triangle? [1:2]
   (2) length of rectangle to length of original triangle? [1:2]
   (3) area of rectangle to area of original triangle? [1:2]

d. If the base of the original triangle is **b** and the height is **h**,
   (1) what is the base of the rectangle? [1/2 **b**]
   (2) what is the height of the rectangle? [1/2 **h**]
   (3) what is the area of the rectangle? [1/4 **bh**]

e. Note that the vertices of the original triangle lay side-by-side at the mid-point of its base. What is their sum? [180°]

Student Record

f. Have the students label their Amazing Circles as shown in the diagram. Instruct the students to color in or shade the rectangle. Once completed, have them roll-tape their Amazing Circle to their page.

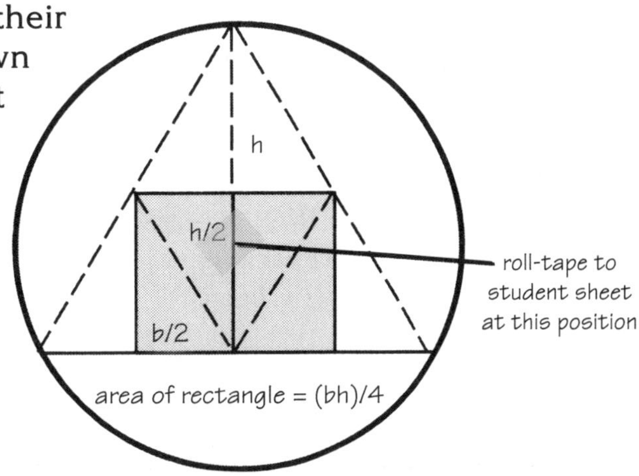

Name Class Date

Amazing Circle Number 12

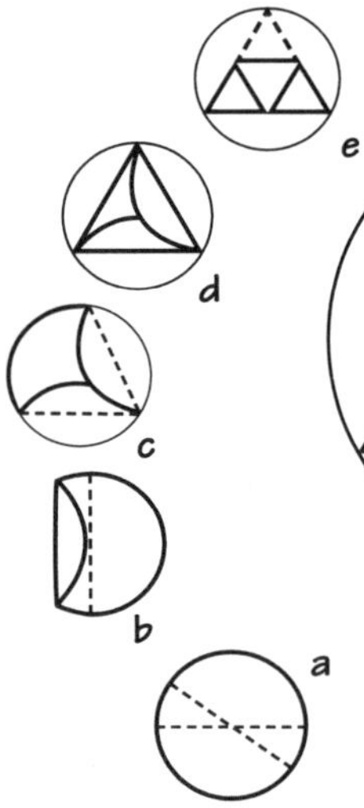

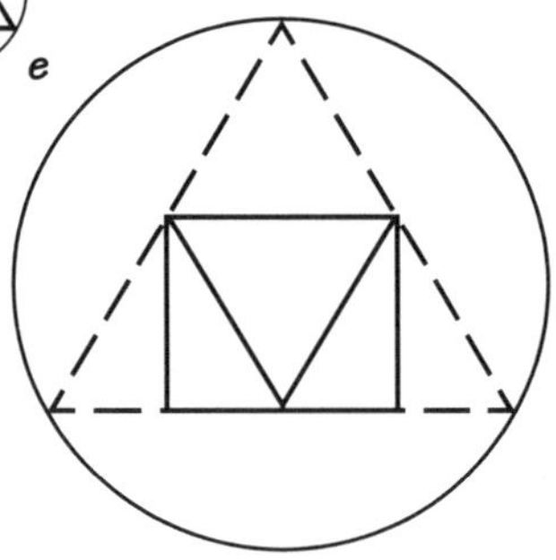

Cut out an Amazing Circle and fold to the isosceles trapezoid (a, b, c, d, e). Fold the second and third vertices of the triangle to the same mid-point.

Identification

a. What type of figure is formed?

Relationships and Measurements

b. What can you say about the two pairs of opposite sides?

c. What are the ratios of the following:
   (1) height of rectangle to height of original triangle?
   (2) length of rectangle to length of original triangle?
   (3) area of rectangle to area of original triangle?

d. If the base of the original triangle is b and the height is h,
   (1) what is the base of the rectangle?
   (2) what is the height of the rectangle?
   (3) what is the area of the rectangle?

roll-tape here

e. Note that the vertices of the original triangle lay side-by-side at the mid-point of its base. What is their sum?

f. Tape your labeled and colored-in Amazing Circle to the outlined circular region.

Exploring the *Amazing Circle* 13

Distribute the student page, *Amazing Circle Number 13.*

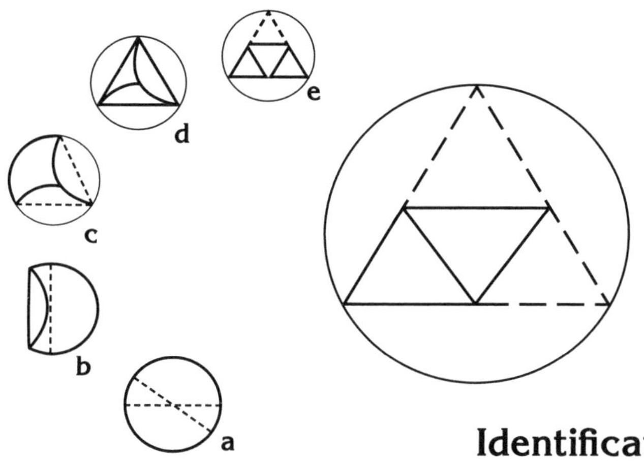

Fold an Amazing Circle to the isosceles trapezoid (a, b, c, d, e). Fold one of the outside triangles over the center triangle.

Identification

a. What is the new figure called? [*Rhombus.* A rhombus is a subset of the set of parallelograms.]

Relationships and Measurements

b. How do the measures of the sides of the rhombus compare? [They are congruent.]
c. What can you say about the pairs of opposite sides? (Parallel and congruent)
d. How do opposite angles compare? [Congruent]
e. What are the measures of the angles? [60˚ and 120˚]
f. What is the sum of measures of the four angles? [360˚]
g. How does this compare with the sum of the angles in the previous trapezoid? [The same]
h. What are the ratios of the following areas:
   (1) rhombus to the original triangle? [2:4 or 1:2]
   (2) trapezoid to the rhombus? [3:2]
   (3) rhombus to the small triangle? [2:1]
   (4) rhombus to the trapezoid? [2:3]
i. What are the ratios of the following perimeters:
   (1) rhombus to the original triangle? [4:6 or 2:3]
   (2) rhombus to the trapezoid? [4:5]
   (3) original triangle to the rhombus? [6:4 or 3:2]
   (4) trapezoid to the rhombus? [5:4]

Student Record

j. Have the students label their Amazing Circles as shown in the diagram. Instruct the students to color in or shade the rhombus. Once completed, have them roll-tape their Amazing Circle to their page.

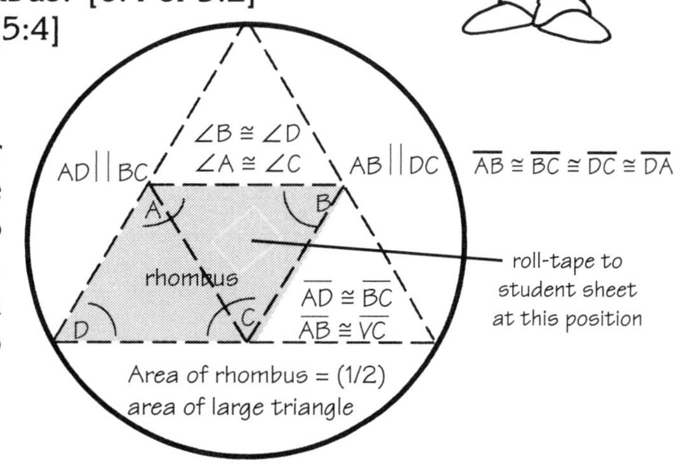

TEACHER GUIDELINES

Name Class Date

Amazing Circle Number 13

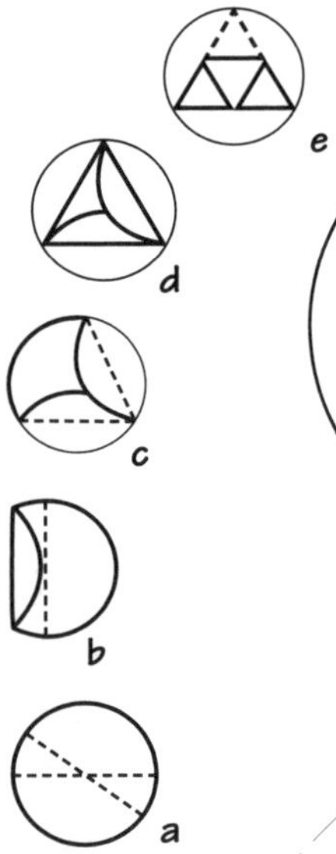

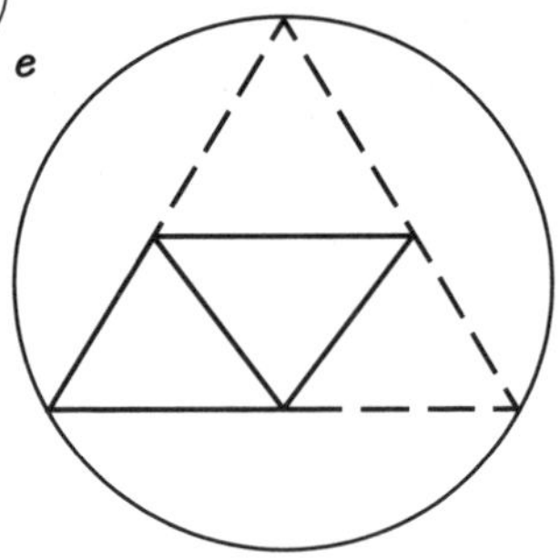

Cut out an Amazing Circle. Fold to the isosceles trapezoid (a, b, c, d, e). Fold one of the outside triangles over the center triangle.

Identification

a. What is the new figure called?

Relationships and Measurements

b. How do the measures of the sides of the rhombus compare?

c. What can you say about the pairs of opposite sides?

d. How do opposite angles compare?

e. What are the measures of the angles?

f. What is the sum of the measures of the four angles?

roll-tape here

g. How does this compare with the sum of the angles in the previous trapezoid?

h. What are the ratios of the following areas:
(1) rhombus to the original triangle?
(2) trapezoid to the rhombus?
(3) rhombus to the small triangle?
(4) rhombus to the trapezoid?

i. What are the ratios of the following perimeters:
(1) rhombus to the original triangle?
(2) rhombus to the trapezoid?
(3) original triangle to the rhombus?
(4) trapezoid to the rhombus?

j. Tape your labeled and colored-in Amazing Circle to the outlined circular region.

Exploring the *Amazing Circle* 14

Distribute the student page, *Amazing Circle Number 14.*

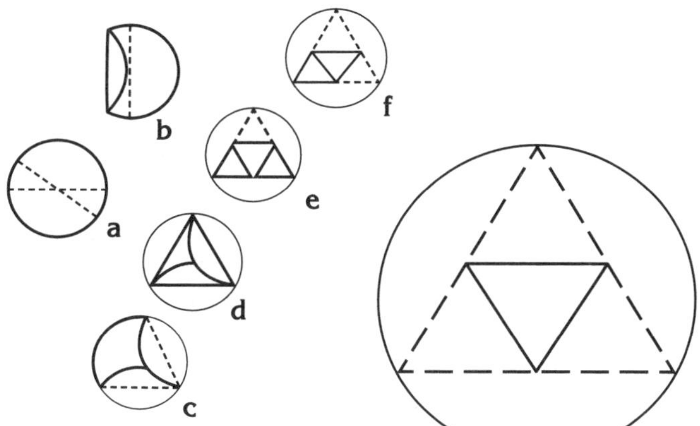

Fold an Amazing Circle to the rhombus (a-f). Fold the remaining outside triangle over the center triangle.

Identification

a. What is the new figure called? [Equilateral triangle]

Relationships and Measurements

b. What are the ratios of the following areas:
   (1) small triangle to the original triangle? [1:4]
   (2) small triangle to the trapezoid? [1:3]
   (3) small triangle to the rhombus? [1:2]
   (4) small triangle to the rectangle? [1:2. The area of the small rectangle is 1/4 that of the original triangle while the area of the rectangle is 1/2 that of the original triangle.]

c. What are the ratios of the following perimeters:
   (1) original triangle to the small triangle? [6:3 or 2:1]
   (2) small triangle to the trapezoid? [3:5]
   (3) small triangle to the rhombus? [3:4]
   (4) trapezoid to the small triangle? [5:3]

Student Record

d. Have the students label their Amazing Circles as shown in the diagram. Instruct the students to color in or shade the central equilateral triangle. Also, instruct them to outline the rhombus, large triangle, rectangle, and trapezoid in different colors and make a color key. Once completed, have them roll-tape their Amazing Circle to their page.

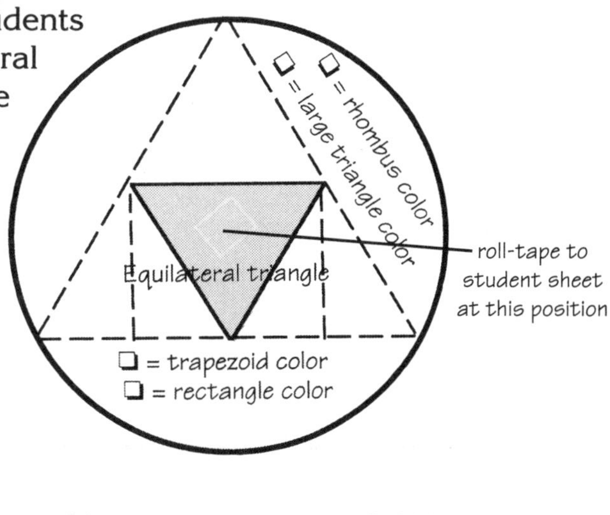

TEACHER GUIDELINES

Name Class Date

Amazing Circle Number 14

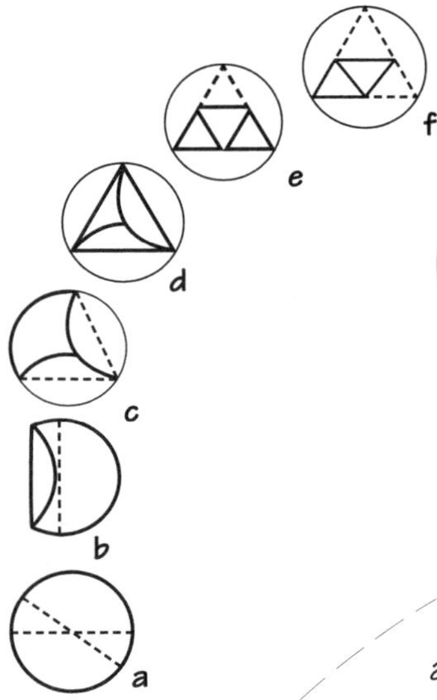

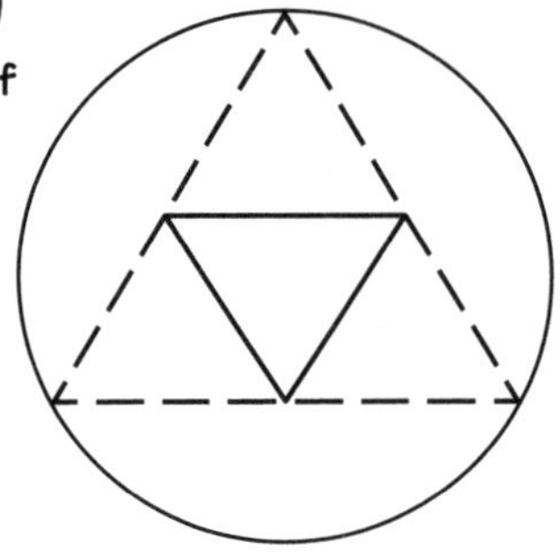

Cut out and fold an Amazing Circle to the rhombus (a-f). Fold the remaining outside triangle over the center triangle.

Identification

a. What is the new figure called?

Relationships and Measurements

b. What are the ratios of the following areas:
   (1) small triangle to the original triangle?

   (2) small triangle to the trapezoid?

   (3) small triangle to the rhombus?

   (4) small triangle to the rectangle?

roll-tape here

c. What are the ratios of the following perimeters:
   (1) original triangle to the small triangle?

   (2) small triangle to the trapezoid?

   (3) small triangle to the rhombus?

   (4) trapezoid to the small triangle?

d. Tape your labeled and colored-in Amazing Circle to the outlined circular region.

Exploring the *Amazing Circle* 15

Distribute the student page, *Amazing Circle Number 15.*

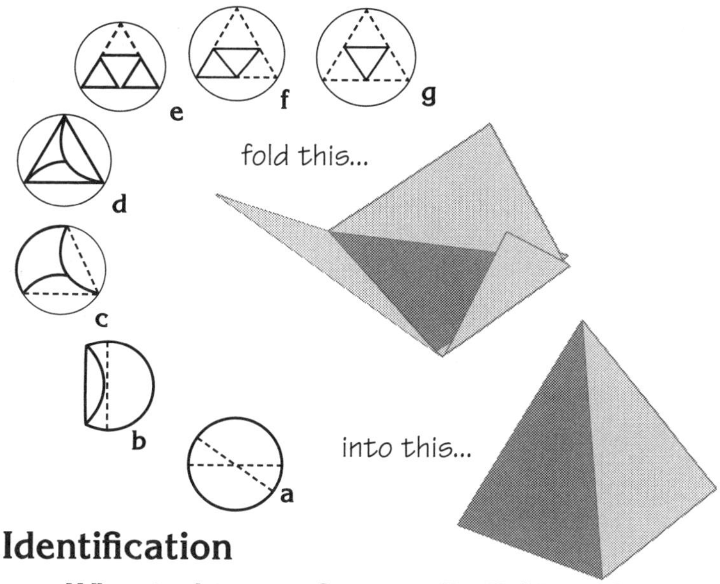

Fold an Amazing Circle to the equilateral triangle (a-d). Fold to the small equilateral triangle (e, f, g). Open the figure to the original triangle. Using the center triangle as a base, fold up the remaining three small triangles to form a three-dimensional figure.

Identification

a. What is this new figure called? [*Regular tetrahedron. Tetra* means four so tetrahedron is a solid with four *faces*. It is *regular* because all faces are congruent.]

Relationships and Measurements

b. How does the surface area of the tetrahedron compare with:
   (1) the area of the original triangle? [The same]
   (2) the area of the small triangle? [4 times as large]
   (3) the area of the trapezoid? [4/3 as large]
   (4) the area of the rhombus? [Twice as large]

c. How many vertices does the tetrahedron have? [4]

d. How many edges does the tetrahedron have? [6]

Student Record

e. Have the students label their Amazing Circles as shown in the diagram. Instruct the students to color in or shade the central triangle that forms the base of the tetrahedron. Direct them to label the three faces. Once completed, have them roll-tape their Amazing Circle to their page.

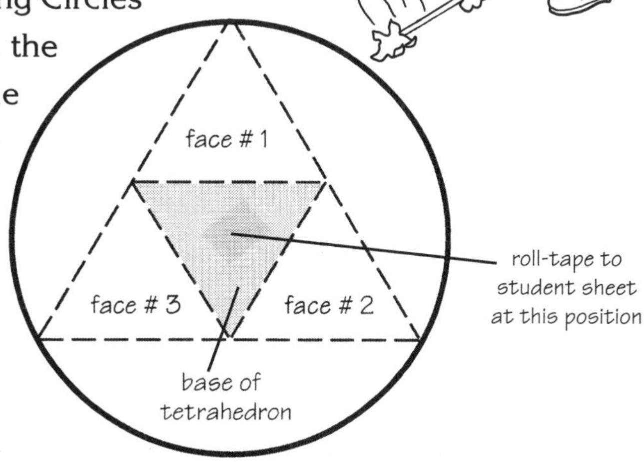

TEACHER GUIDELINES

Name Class Date

Amazing Circle Number 15

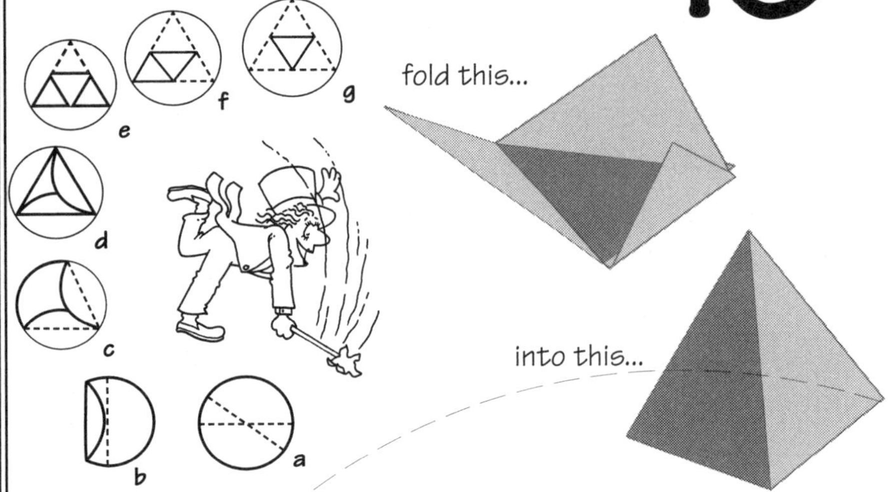

Cut out an Amazing Circle and fold to the equilateral triangle (a-d). Fold again to the small equilateral triangle (e, f, g). Open the figure to the original triangle. Using the center triangle as a base, fold up the remaining three small triangles to form a three-dimensional figure.

Identification

a. What is this new figure called?

Relationships and Measurements

b. How does the surface area of the tetrahedron compare with:

(1) the area of the original triangle?

(2) the area of the small triangle?

(3) the area of the trapezoid?

(4) the area of the rhombus?

c. How many vertices does the tetrahedron have?

roll-tape here

d. How many edges does the tetrahedron have?

e. Tape your labeled and colored-in Amazing Circle to the outlined circular region.

Exploring the *Amazing Circle* 16

Distribute the student page, *Amazing Circle Number 16.*

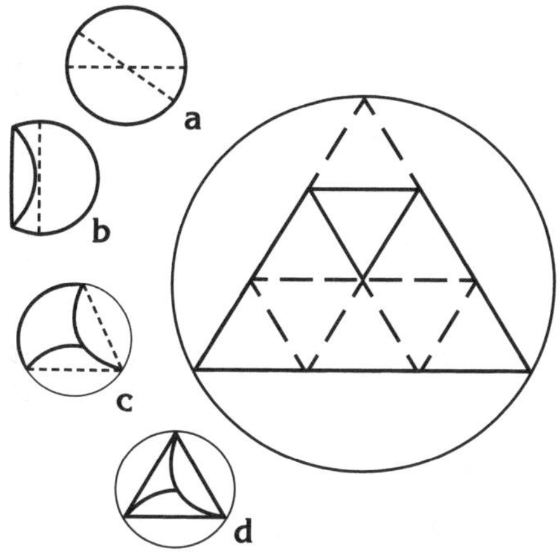

Fold the Amazing Circle to the equilateral triangle (a-d). Fold a vertex to the center of the circle. Trace around the two sides of the triangle. Open to the original triangle. Highlight the crease line. Repeat for the other two vertices. Open the figure to the original triangle. The triangle has now been divided into nine smaller congruent triangles. Fold one vertex back to the center of the circle.

Identification

a. What is a name for the new figure? [Isosceles trapezoid]

Relationships and Measurements

b. How does the area of this trapezoid compare with that of the original triangle? [8/9 as large]

c. How does the perimeter compare with that of the original triangle? [8/9 as long]

d. How does the short base of the trapezoid compare with:
   (1) the long base? [1/3 as long]
   (2) one of the non-parallel sides? [1/2 as long]

e. How does the area of this new and smallest triangle compare with that of the original triangle? [1/9 as large]

Student Record

f. Have the students label their Amazing Circles as shown in the diagram. Instruct them to number the triangles as shown. Direct them to outline the isosceles trapezoid in one color, its perimeter in a different color, and the perimeter of the large triangle in a third color. They can also number the sides of each triangle that is part of a perimeter. Once completed, have them roll-tape their Amazing Circle to their page.

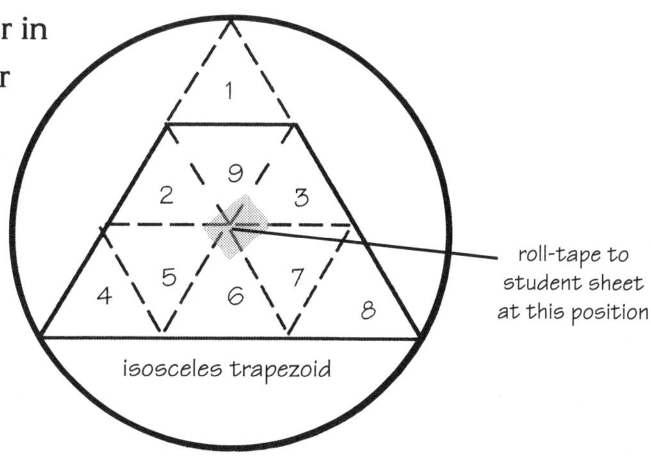

TEACHER GUIDELINES

Name Class Date

Amazing Circle Number 16

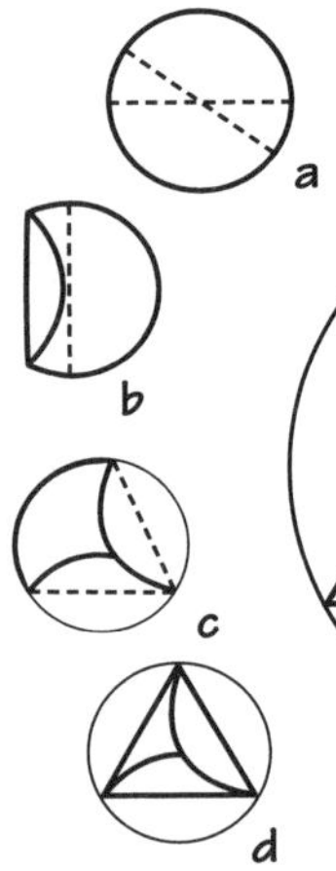

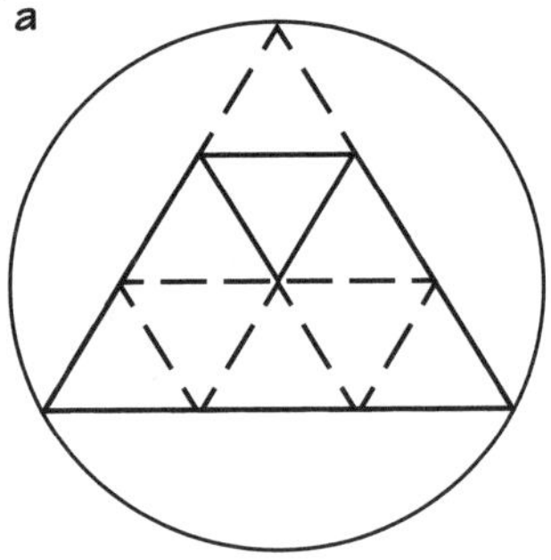

Cut out an Amazing Circle and fold to the equilateral triangle (a-d). Fold a vertex to the center of the circle. Trace around the two sides of the triangle. Open to the original triangle. Highlight the crease line. Repeat for the other two vertices. Open the figure to the original triangle. The triangle has now been divided into nine smaller, congruent triangles. Fold one vertex back to the center of the circle.

Identification

a. What is a name for the new figure?

Relationships and Measurements

b. How does the area of this trapezoid compare with that of the original triangle?

c. How does the perimeter compare with that of the original triangle?

roll-tape here

d. How does the short base of the trapezoid compare with:
(1) the long base?

(2) one of the non-parallel sides?

e. How does the area of this new and smallest triangle compare with that of the original triangle?

f. Tape your labeled and colored-in Amazing Circle to the outlined circular region.

Exploring the *Amazing Circle* 17

Distribute the student page, *Amazing Circle Number 17.*

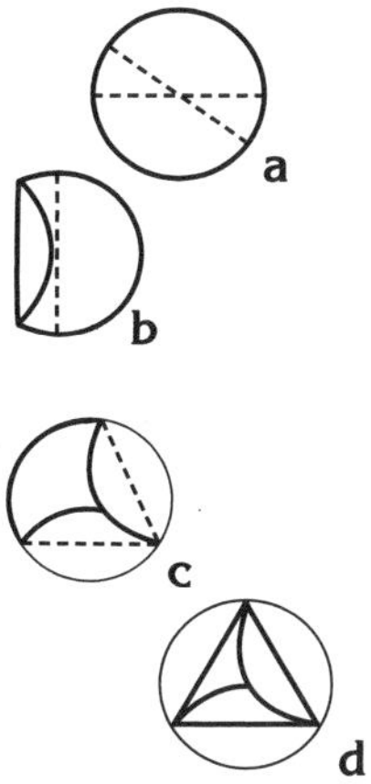

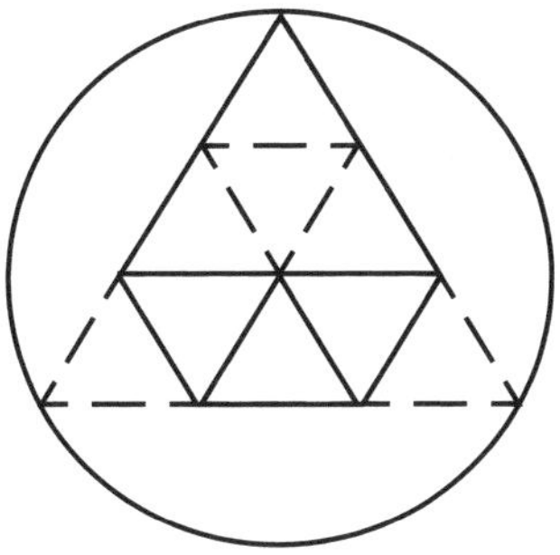

Fold an Amazing Circle to the equilateral triangle (a-d). Fold a vertex to the center of the circle. Fold a second vertex to the center.

TEACHER GUIDELINES

Identification

a. What is a name for this new figure? [*Pentagon*]

Relationships and Measurements

b. What are the area ratios of the following:
   (1) pentagon to the original triangle? [7:9]
   (2) pentagon to the preceding isosceles trapezoid? [7:8]

c. What are the perimeter ratios of the following:
   (1) pentagon to the original triangle? [7:9]
   (2) pentagon to the preceding isosceles trapezoid? [7:8]

Student Record

d. Have the students label their Amazing Circles as shown in the diagram. Instruct the students to number the triangles as shown. Have them outline the perimeter of the larger triangle in one color, and the perimeter of the pentagon in a different color. They can also number the sides of each triangle that is part of a perimeter. Once completed, have them roll-tape their Amazing Circle to their page.

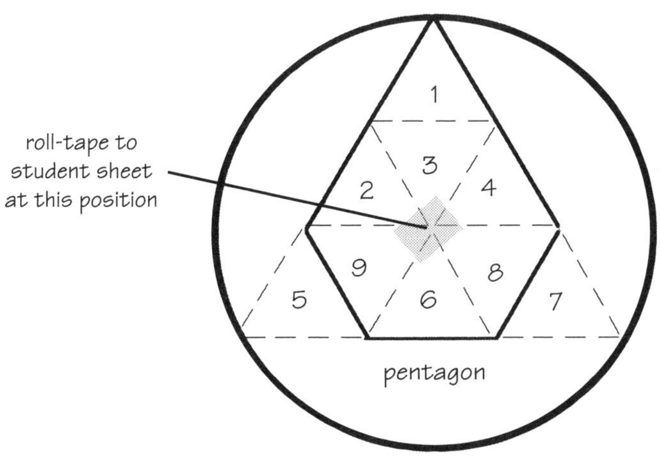

Name Class Date

Amazing Circle Number 17

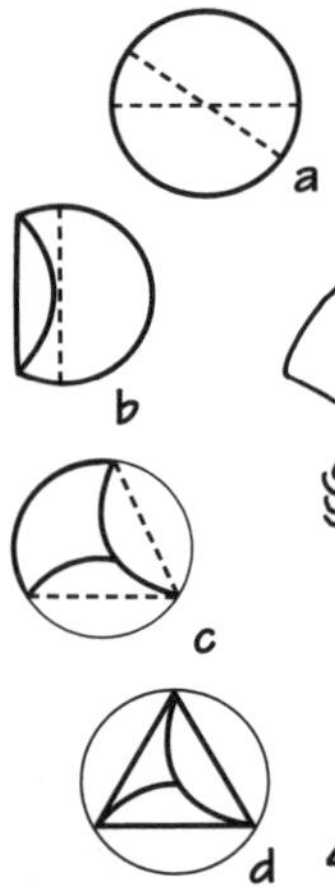

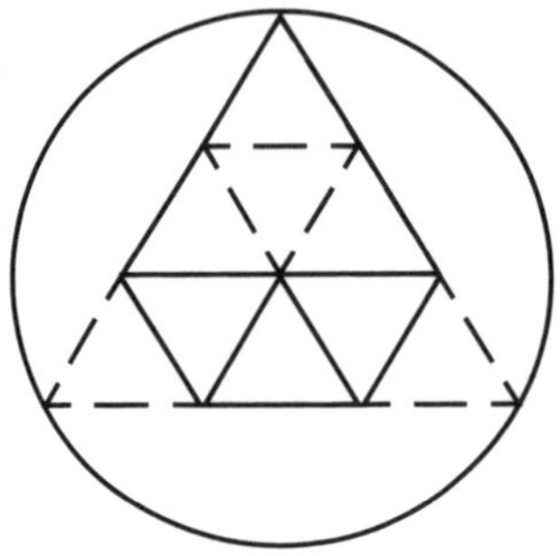

Cut out an Amazing Circle and fold to the equilateral triangle (a-d). Fold a vertex to the center of the circle. Fold a second vertex to the center.

Identification

a. What is a name for this new figure?

Relationships and Measurements

b. What are the area ratios of the following:
   (1) pentagon to the original triangle?

   (2) pentagon to the preceding isosceles trapezoid?

roll-tape here

c. What are the perimeter ratios of the following:
   (1) pentagon to the original triangle?

   (2) pentagon to the preceding isosceles trapezoid?

d. Tape your labeled and colored-in Amazing Circle to the outlined circular region.

Exploring the *Amazing Circle* 18

Distribute the student page, *Amazing Circle Number 18.*

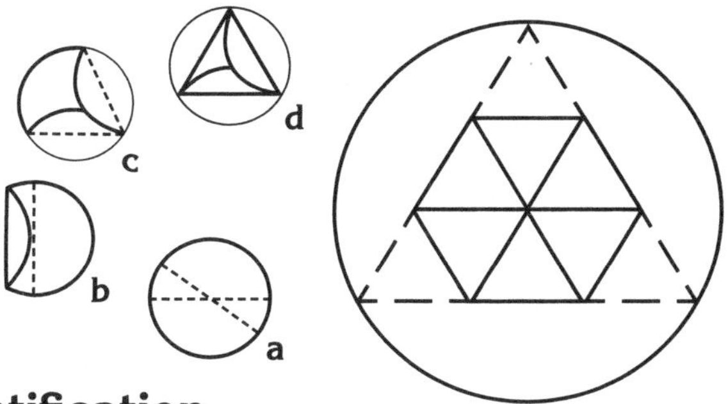

Fold an Amazing Circle to the equilateral triangle (a-d). Fold all three vertices to the center of the circle.

Identification

a. What is a name for this new figure? [*Regular hexagon.* A regular hexagon has all sides and interior angles congruent.]

Relationships and Measurements

b. How many small triangles do you see? [6]
c. How many small triangles are covered up? [3]
d. What are the area ratios of the following:
  (1) hexagon to the original triangle? [6:9 or 2:3]
  (2) hexagon to the pentagon? [6:7]
  (3) hexagon to the trapezoid in this sequence? [6:8 or 3:4]
e. What are the ratios of the perimeters in the following:
  (a) hexagon to the original triangle? [6:9 or 2:3]
  (b) hexagon to the pentagon? [6:7]
  (c) hexagon to the trapezoid in this sequence? [6:8 or 3:4]
f. How does the area of the hexagon compare with that of the first trapezoid, not the one in this sequence? [The area of the hexagon is 2/3 or 8/12 that of the original triangle and the area of the trapezoid is 3/4 or 9/12 that of the triangle. Therefore, the area of the hexagon is 8/9 that of the trapezoid.]
g. What are the ratios of the areas in the following:
  (1) hexagon to the rhombus? [8:6 or 4:3]
  (2) hexagon to the small triangle in the first series? [8:3]
  (3) hexagon to the small triangle in this series? [6:1]

Student Record

h. Have the students label their Amazing Circles as shown in the diagram. Instruct the students to number the triangles as shown. Have them use a different color to outline each of the following perimeters:
  (1) the larger triangle,
  (2) the pentagon,
  (3) the trapezoid,
  (4) the hexagon.
  Direct them to include a color key. They can also number the sides of each triangle that is part of a perimeter. Once completed, have them roll-tape their Amazing Circle to their page.

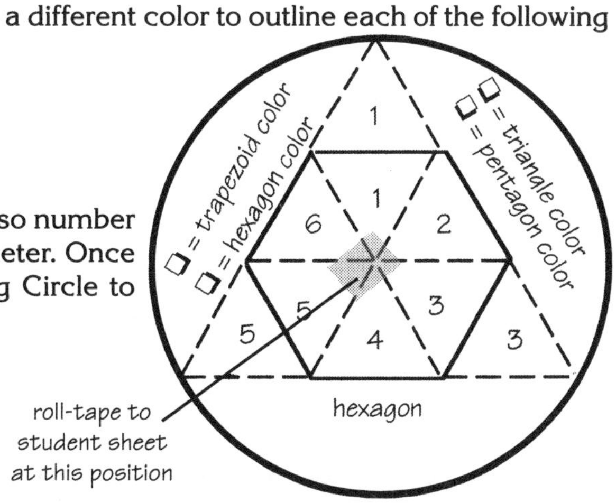

It is clear at this stage that additional rather difficult ratios could be examined. These are left to the discretion of the teacher to create and assign.

TEACHER GUIDELINES

Name Class Date

Amazing Circle Number 18

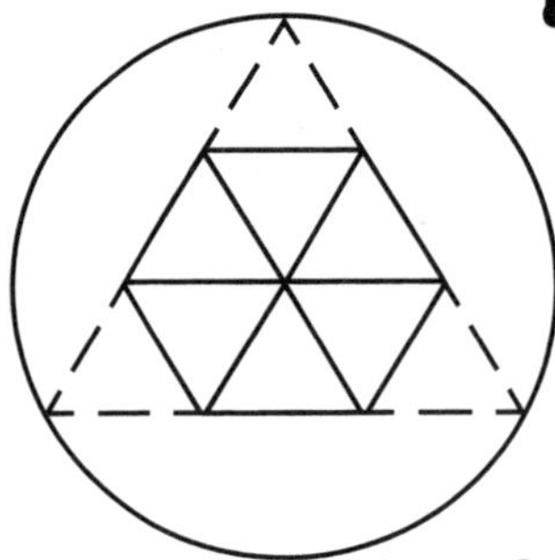

Cut out an Amazing Circle and fold to the equilateral triangle (a-d). Fold all three vertices to the center of the circle.

Identification

a. What is a name for this new figure?

Relationships and Measurements

b. How many small triangles do you see?

c. How many small triangles are covered up?

d. What are the area ratios of the following:
  (1) hexagon to the original triangle?
  (2) hexagon to the pentagon?
  (3) hexagon to the trapezoid in this sequence?

e. What are the ratios of the perimeters in the following:
  (a) hexagon to the original triangle?
  (b) hexagon to the pentagon?
  (c) hexagon to the trapezoid in this sequence?

roll-tape here

f. How does the area of the hexagon compare with that of the first trapezoid, not the one in this sequence?

g. What are the ratios of the areas in the following:
  (1) hexagon to the rhombus?
  (2) hexagon to the small triangle in the first series?
  (3) hexagon to the small triangle in this series?

h. Tape your labeled and colored-in Amazing Circle to the outlined circular region.

Exploring the *Amazing Circle* 19

Distribute the student page, *Amazing Circle Number 19.*

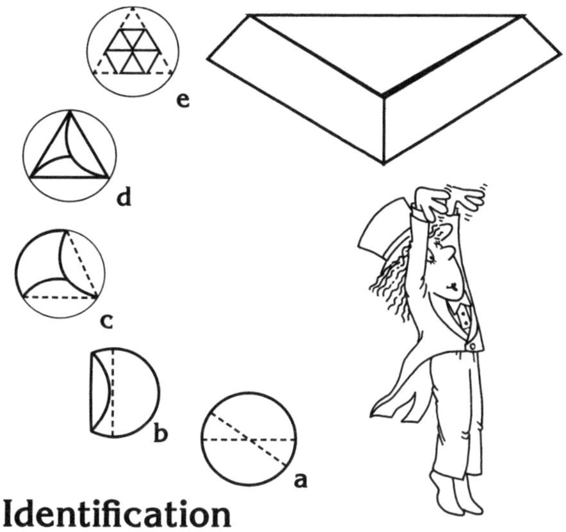

Fold the Amazing Circle to the large equilateral triangle (a-d). Fold to the small equilateral triangle (e). Open up the triangle. Tuck one of the small triangles at one of the vertices into the small triangle at one of the other vertices. Then tuck the triangle at the remaining vertex underneath the other to form a three-dimensional figure.

Identification

a. What is the new figure called? [It is a *truncated tetrahedron.* Truncated means it has a part that has been cut off.]

Relationships and Measurements

b. If the original triangle has an area of 1, what is the area of:
   (1) the bottom base? [1/4]
   (2) the top base? [1/9]
   (3) a side face or wall? [1/4 - 1/9. Note that together one side face and the top base are congruent to the bottom base.]

c. What is the total surface area of the truncated tetrahedron? [[1/4 + 1/9 +3(1/4 - 1/9)] = [9/36 + 4/36 + 3(9/36 - 4/36)] = [13/36 + 3(5/36)] = 28/36 = 7/9]

d. The famous mathematician and "Father of the Problem-Solving Movement," George Polya, advised "It is better to work one problem five ways than to work five problems one way." Can you think of another way to solve this problem? [When the truncated tetrahedron was formed, we hid exactly two of the small triangles. All the rest of the original triangle is to be found in the bases and side faces. Each of the hidden triangles had 1/9 the area of the original triangle. Therefore, 2/9 of that area was lost, leaving 7/9.]

Student Record

e. Have the students label their Amazing Circles as shown in the diagram. Instruct the students to number the triangles as shown. Direct them to then draw in the crease lines that produced the truncated tetrahedron. Once completed, have them roll-tape their Amazing Circle to their page.

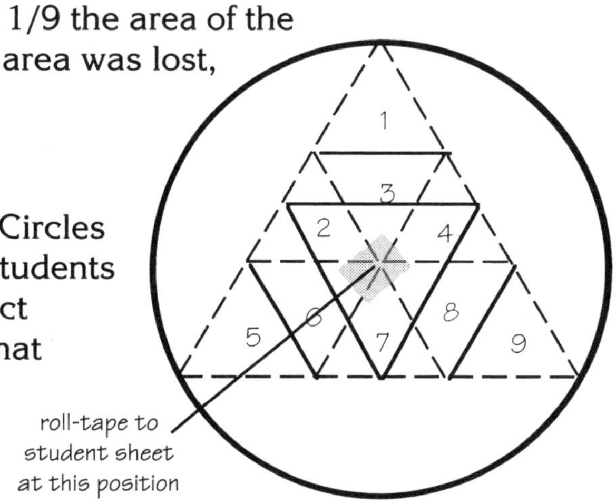

TEACHER GUIDELINES

Name Class Date

Amazing Circle Number 19

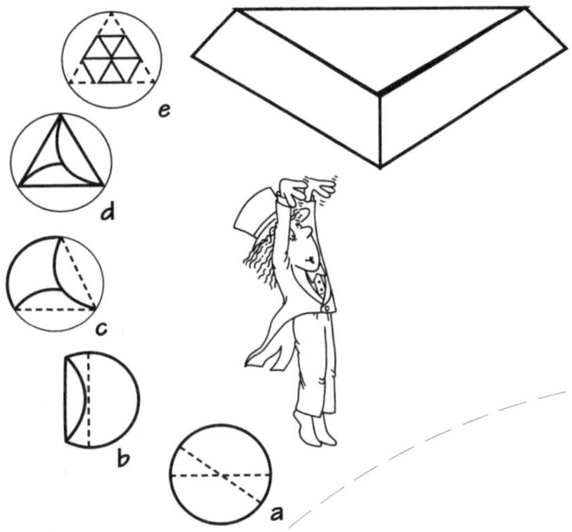

Fold the Amazing Circle to the large equilateral triangle (a-d). Fold to the small equilateral triangle (e). Open up the triangle. Tuck one of the small triangles at one of the vertices into the small triangle at one of the other vertices. Then tuck the triangle at the remaining vertex underneath the other to form a three-dimensional figure.

Identification

a. What is the new figure called?

Relationships and Measurements

b. If the original triangle has an area of 1, what is the area of:

(1) the bottom base?

(2) the top base?

(3) a side face or wall?

c. What is the total surface area of the truncated tetrahedron?

roll-tape here

d. Can you think of another way to solve this problem? Explain.

e. Tape your labeled and colored-in Amazing Circle to the outlined circular region.

Exploring the *Amazing Circle* 20

Distribute the student page, *Amazing Circle Number 20.*

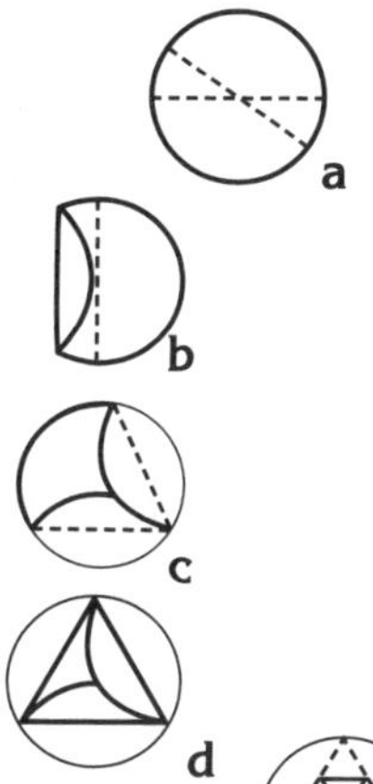

Now that you have constructed your first truncated tetrahedron (a-f), your next challenge is to build a set of three similar truncated tetrahedra that, when stacked, form another taller truncated tetrahedron.

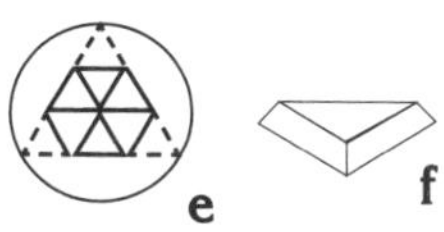

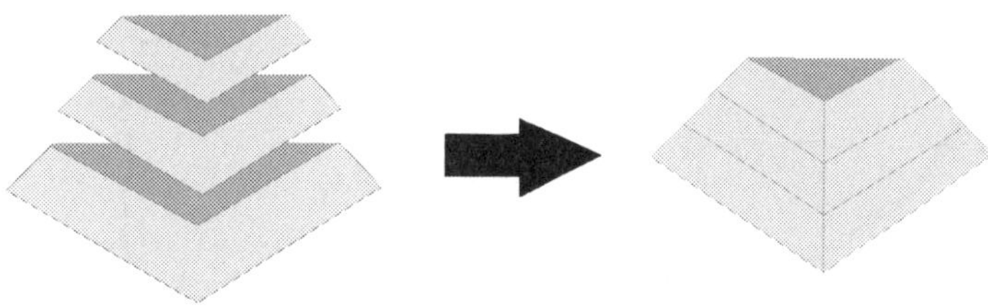

You can store all three in one package by nesting the smallest inside the mid-sized one and the mid-sized one in the largest.

Your first step is to determine the sizes of the circles that will fold into the tetrahedra meeting these conditions. Please describe how you solved this problem. Show and discuss any computations you used to solve the problem.

A geometric approach to this problem is as follows: Consider an Amazing Circle folded to triangle ABC and then to triangle DEF.

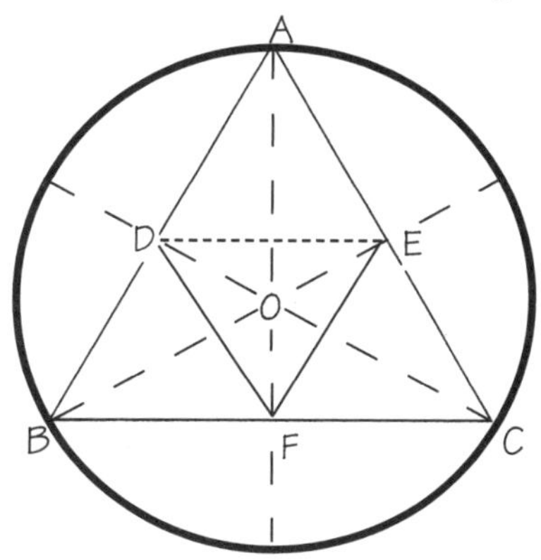

Take triangle DEF as the *base* of the bottom truncated tetrahedron. Folding points A, B, and C to the center point O creates nine small, congruent, equilateral triangles.

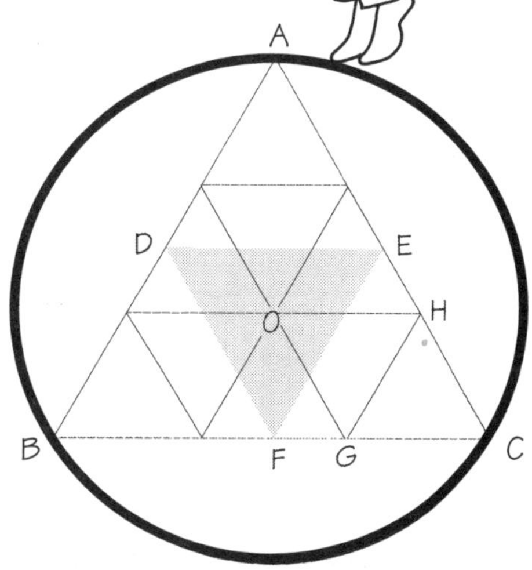

TEACHER GUIDELINES

The small equilateral triangles at the corners of triangle ABC (for example, triangle CGH) fold over to form the *top* of the bottom truncated tetrahedron. Triangle CGH is then the *base* of the middle truncated tetrahedron.

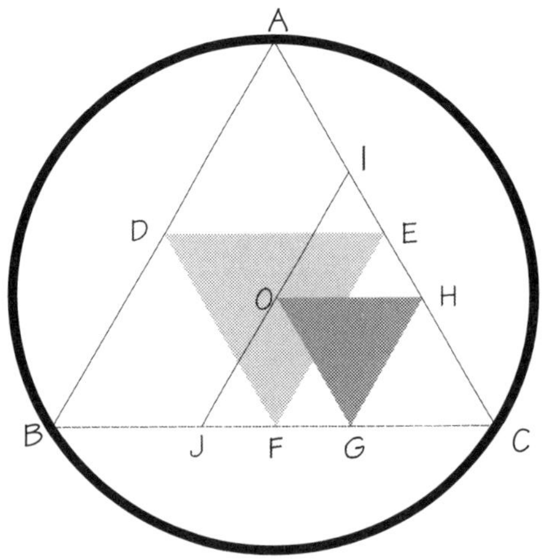

But triangle CGH is congruent to triangle OHG. Therefore, triangle IJC inscribed in a circle builds the next truncated tetrahedron. The center of this inscribed triangle is found by constructing the altitudes IG, OC, and JH. Their intersection is M, the center of the circle that inscribes triangle IJC.

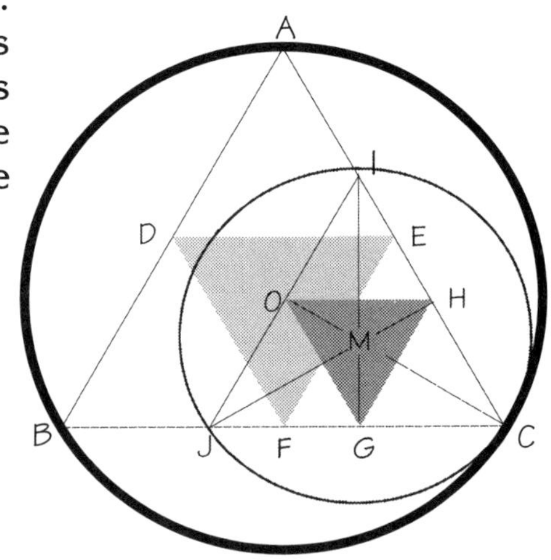

Triangle IJC can also be divided into nine congruent, equilateral triangles. Triangle MKP is the *base* of the *top* truncated tetrahedron. The radius of the circle that inscribes triangle EFC is QN.

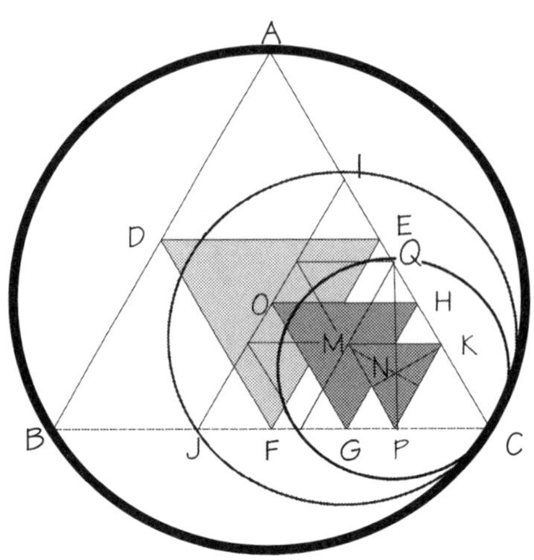

The Amazing Circles for building the set of three similar truncated tetrahedrons have now been nested in the same circle. A page-size pattern (suitable for duplication) for building the set of three nested truncated tetrahedrons follows on the next page.

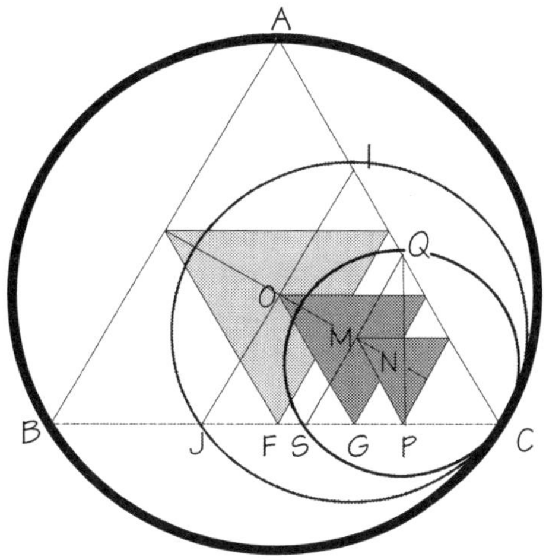

A computational approach considers that all equilateral triangles are similar and, since corresponding lines of similar triangles are in proportion, their radii are in proportion. Let r equal the radius of the largest circle (center at point O). Then the radius of the next circle (center at point M) is 2/3(r). The radius of the smallest circle (center at N) is 2/3[2/3(r)].

A scale for setting the radius of a compass also appears on the page. Use the scale to verify the computational result (r, 2/3(r), 2/3[2/3(r)]) for r=12 units. The scale produces larger nested tetrahedra than previously folded.

TEACHER GUIDELINES

Make three copies of this page. Cut and fold each of the Amazing Circles to build the nested set of truncated tetrahedrons. Triangle DEF is the base of the bottom truncated tetrahedron. Triangle OHG is the base of the middle truncated tetrahedron, and triangle MKP is the base of the top truncated tetrahedron.

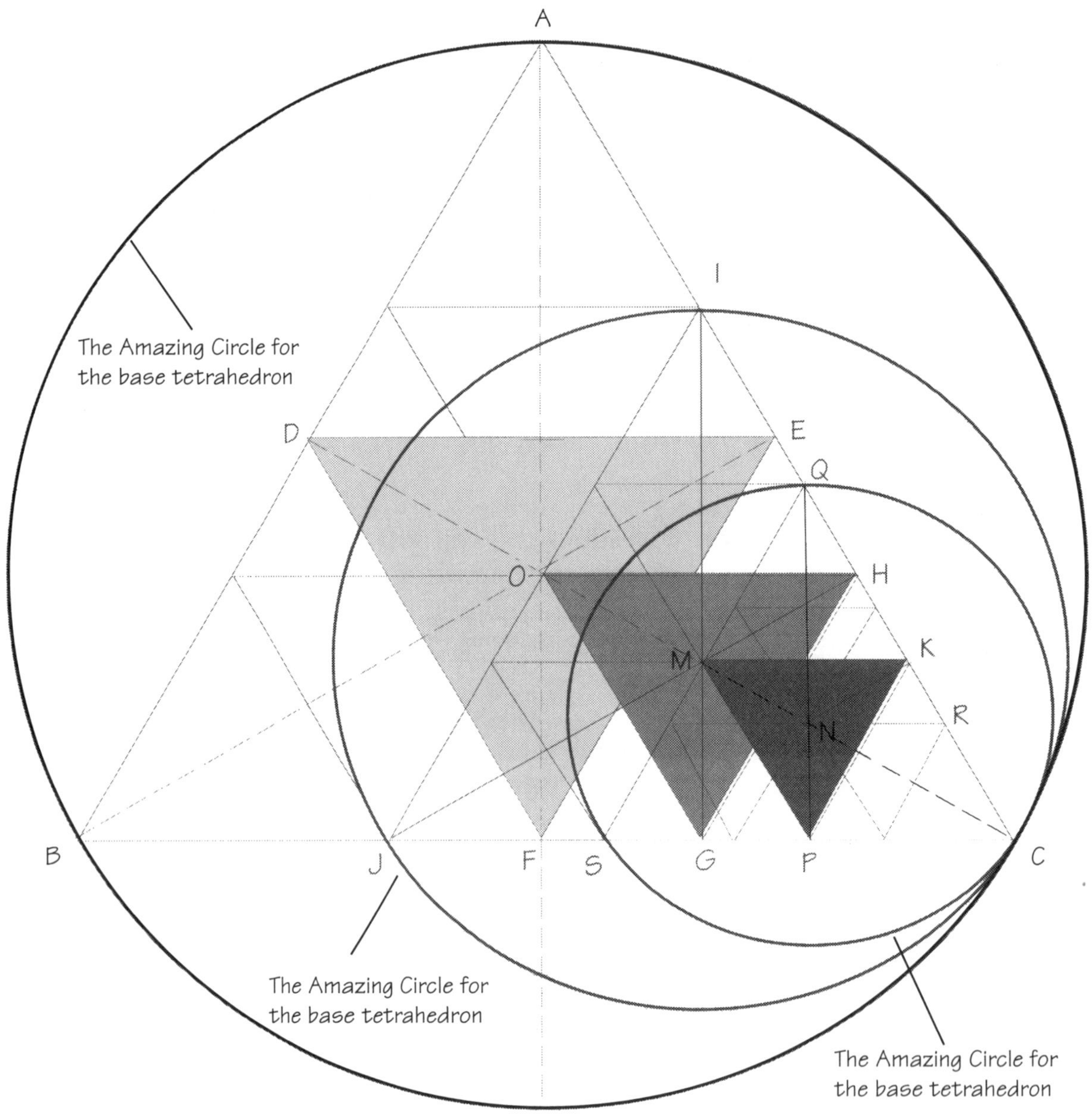

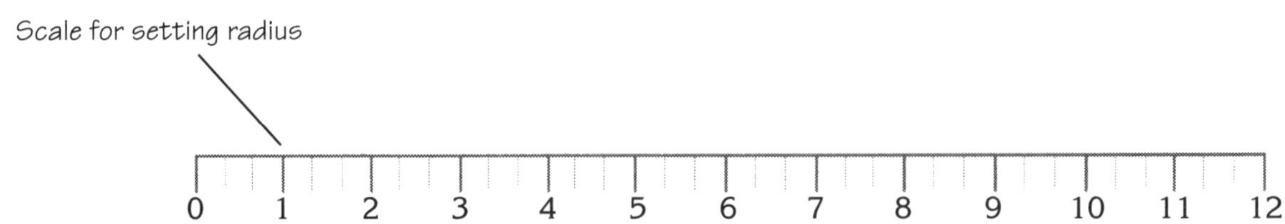

Name Class Date

Amazing Circle Number 20

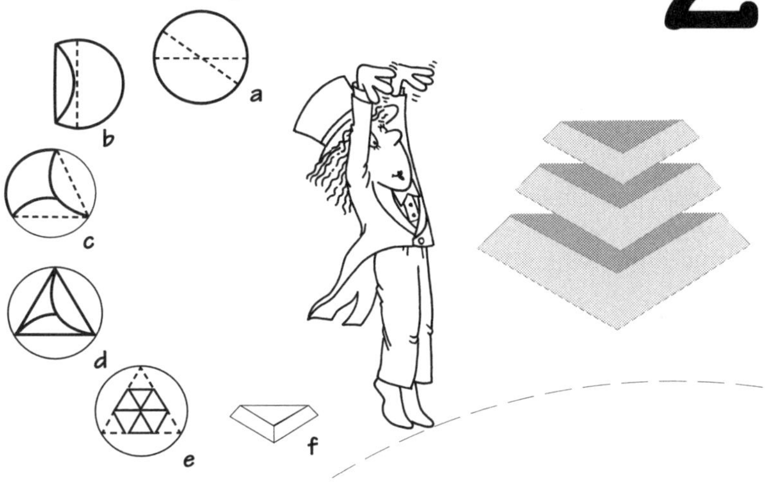

Now that you have constructed your first truncated tetrahedron (a-f), your next challenge is to build a set of three similar truncated tetrahedra that, when stacked, form another taller truncated tetrahedron.

You can store all three in one package by nesting the smallest inside the mid-sized one and the mid-sized one in the largest.

Your first step is to determine the sizes of the circles that will fold into the tetrahedra meeting these conditions. Please describe how you solved this problem. Show and discuss any computations you used to solve the problem.

Exploring the *Amazing Circle* 21

Distribute the student page, *Amazing Circle Number 21.*

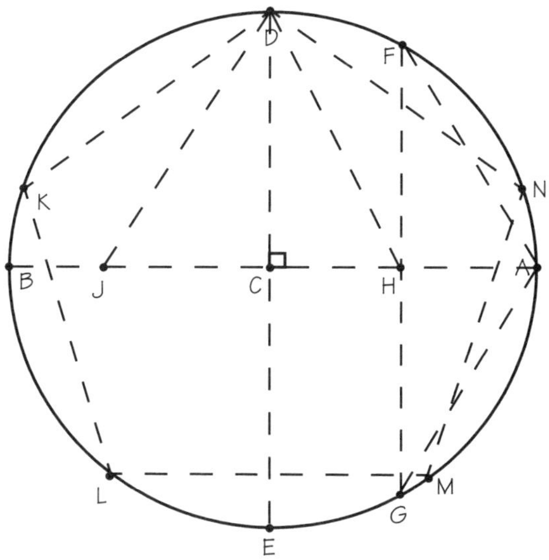

Instructions for folding The Amazing Circle into an equilateral triangle and a square are explained in The Amazing Circle, numbers 7 and 5. For completing the folds required to fold the faces of the Platonic solids, the instructions for folding an Amazing Circle into a regular pentagon are as follows:

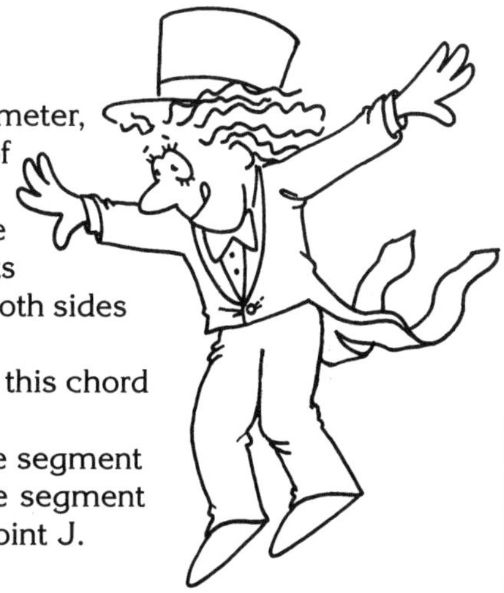

a. Cut out an Amazing Circle.
b. Fold a diameter with end points A and B. Fold a second diameter, with endpoints D and E, that is the perpendicular bisector of the first diameter. Label the center of the circle C.
c. The line segment from point A to point C is a radius of the circle. Fold the circle in half along the diameter with endpoints A and B. Keep point A fixed and fold point C to the circle, on both sides of the diameter, and label these points F and G.
d. Fold the chord between points F and G. Label the point where this chord intersects the diameter with endpoints A and B, H.
e. Fold the chord connecting points F and G under. Fold the line segment with endpoints H and D. Keep point H fixed and fold the line segment $\overline{DH}$ onto the diameter with endpoints A and B to locate the point J.
f. Fold the line segment connecting points D and J.
g. Keep point D fixed and fold the line segment connecting points D and J so that the point J is on the circle. Label this point K (when done properly, a crease line will bisect angle KDJ).
h. Fold the chord connecting points D and K. This chord is the side of the regular, inscribed pentagon.
i. Keep point K fixed. Going counter-clockwise, use the chord between points D and K to locate the endpoint of the next side of the pentagon. Label this endpoint L.
j. Keep point L fixed and again going counter-clockwise, use the chord between points K and L to locate the endpoint of the next side of the pentagon. Label this endpoint M.
k. Keep point M fixed. Going counter-clockwise, use the chord between points L and M to locate the endpoint of the next side of the pentagon. Label this endpoint N.
l. Fold the chord between points D and N to finish the regular pentagon.

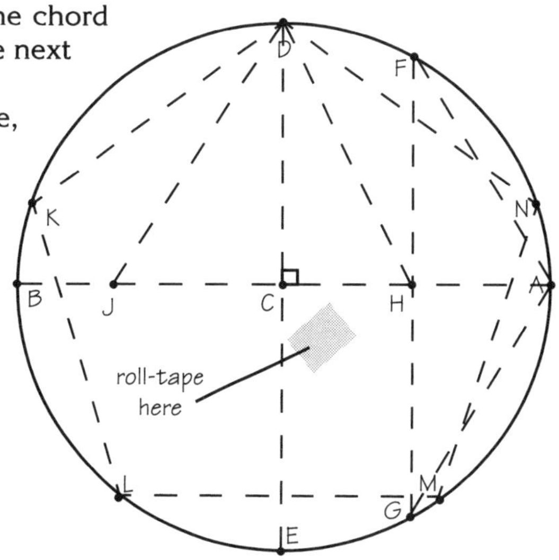

Student Record

m. Direct the students to label their Amazing Circles as shown in the diagram. Once completed, have them roll-tape their Amazing Circle to their student page.

TEACHER GUIDELINES

Name Class Date

Amazing Circle Number 21

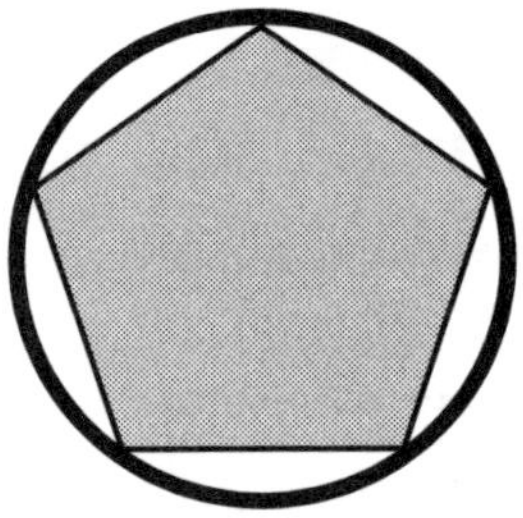

Follow these steps to fold an Amazing Circle into a regular pentagon.

a. Cut out an Amazing Circle.

b. Fold a diameter with end points A and B. Fold a second diameter, with endpoints D and E, that is the perpendicular bisector of the first diameter. Label the center of the circle C.

c. The line segment from point A to point C is a radius of the circle. Fold circle in half along the diameter with endpoints A and B. Keep point A fixed and fold the point C to the circle, on both sides of the diameter, and label these points F and G.

d. Fold the chord between points F and G. Label the point where this chord intersects the diameter with endpoints A and B, H.

e. Fold the chord connecting points F and G under. Fold the line segment with endpoints H and D. Keep point H fixed and fold the line segment $\overline{DH}$ onto the diameter with endpoints A and B to locate the point.

f. Fold the line segment connecting points D and J.

g. Keep point D fixed and fold the line segment connecting points D and J so that the point J is on the circle. Label this point K (when done properly, a crease line will bisect angle KDJ).

h. Fold the chord connecting points D and K. This chord is the side of the regular, inscribed pentagon.

i. Keep point K fixed. Going counter-clockwise, use the chord between points D and K to locate the endpoint of the next side of the pentagon. Label this endpoint L.

j. Keep point L fixed and again going counter-clockwise, use the chord between points K and L to locate the endpoint of the next side of the pentagon. Label this endpoint M.

k. Keep point M fixed. Going counter-clockwise, use the chord between points L and M to locate the endpoint of the next side of the pentagon. Label this endpoint N.

l. Fold the chord between points D and N to finish the regular pentagon.

roll-tape here

Exploring the Amazing Circle In 3-Dimensions

Geometry — The Amazing Circle

The Five Platonic Solids

There are only five solid figures that can be constructed from regular triangles, regular squares, and regular pentagons. Regular polygons with more than five sides, like a hexagon, will not fit together to form solid figures.

Around 400 B. C. Plato, the ancient Greek philosopher and geometer, described in detail these solids, hence their name, the Platonic Solids.

Four equilateral triangles form the faces of a tetrahedron.

Six squares form the faces of a hexahedron.

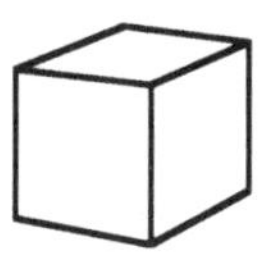

Eight equilateral triangles form the faces of an octahedron.

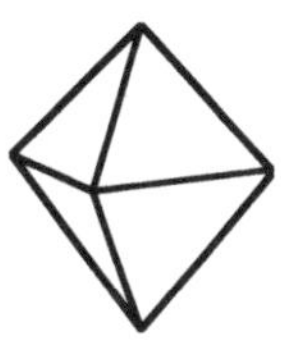

Twelve regular pentagons form the faces of a dodecahedron.

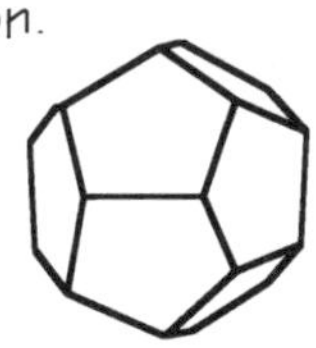

Twenty equilateral triangles form the faces of an icosahedron.

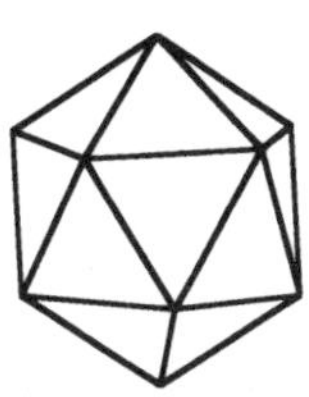

The Five Platonic Solids

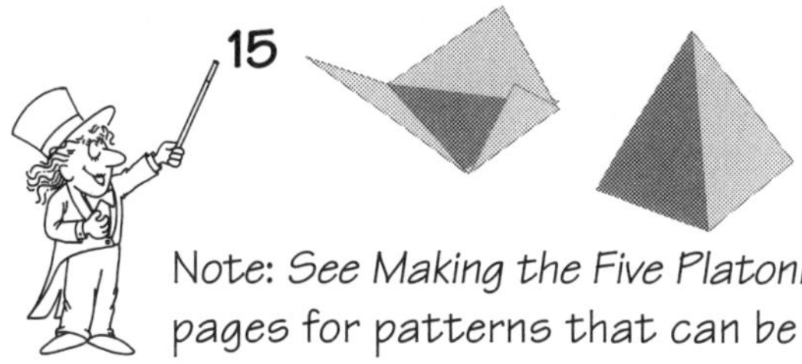

The Amazing Circle folded into a tetrahedron.

Note: *See Making the Five Platonic Solids* on the following pages for patterns that can be duplicated, cut out, folded, and glued.

Tetrahedron

Four Amazing Circles folded to an equilateral triangle make a tetrahedron.

Hexahedron

Six Amazing Circles folded to a square make a hexahedron.

Octahedron

Eight Amazing Circles folded to an equilateral triangle make an octahedron.

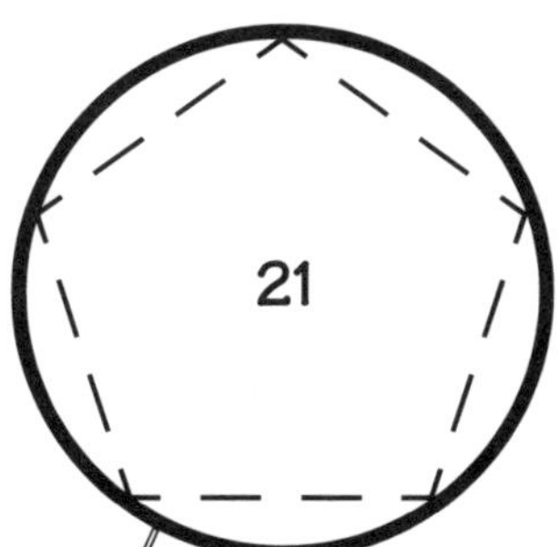

Dodecahedron

Twelve Amazing Circles folded to a regular pentagon make a dodecahedron.

Icosahedron

Twenty Amazing Circles folded to an equilateral triangle make an icosahedron.

Making the Five Platonic Solids

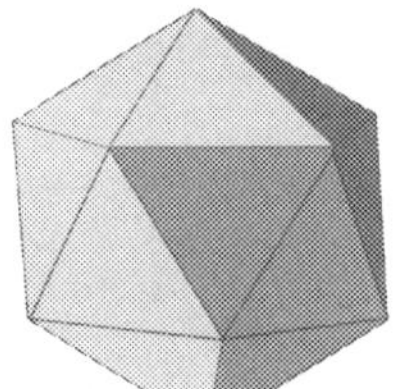

A common method is used to make each of the five Platonic solids. The simplest solid, the tetrahedron, will be used to illustrate the process.

1. Decide which solid to make and either fold circles to the shape for each face required or duplicate the required number of the patterns found on the following pages. (For the tetrahedron, four equilateral triangles are required.)

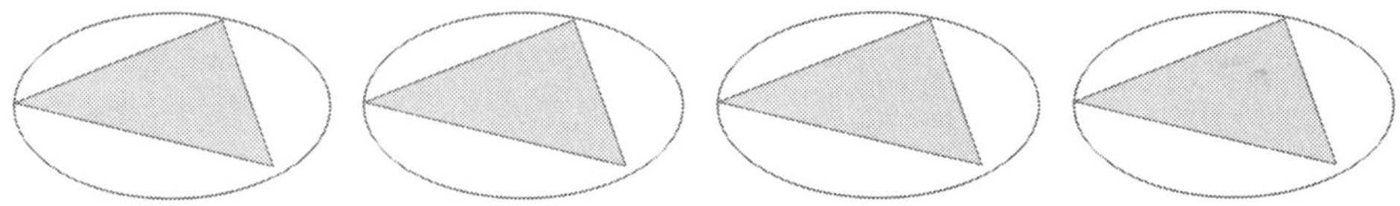

2. To form the flaps for each of the four circles, fold up the chords forming the sides of each equilateral triangle to form flaps.

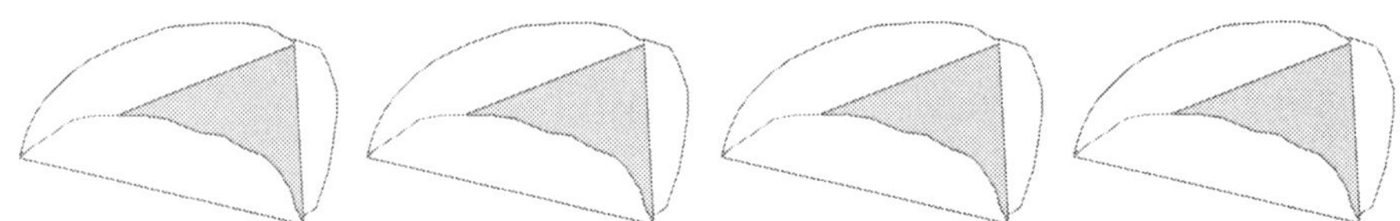

3. Glue the flaps of two circles together.

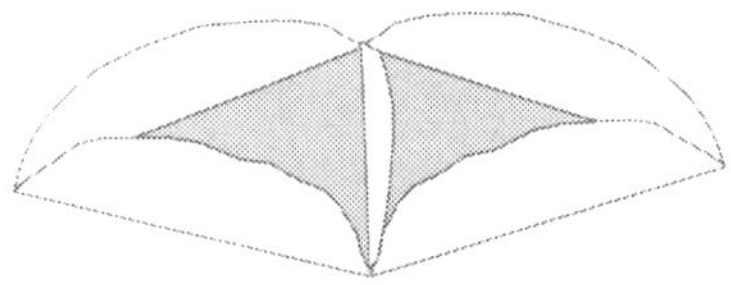

4. Glue the flap of a third circle to a flap of the pair.

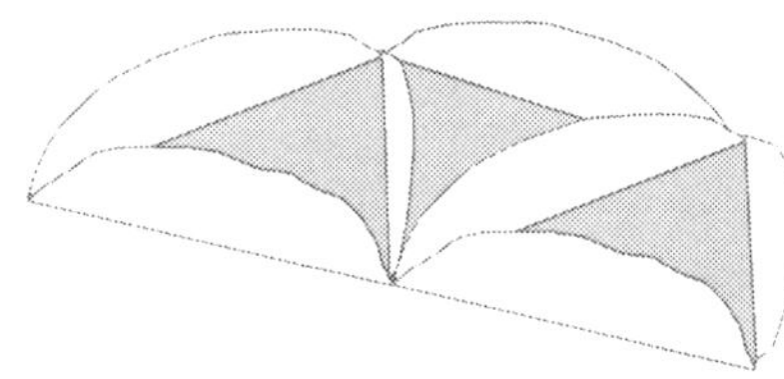

5. Glue the flaps labeled 1 and 2 together to form a triangular-based pyramid. You may choose to glue the flaps together so all of the flaps are on the inside — the easiest — or on the outside.

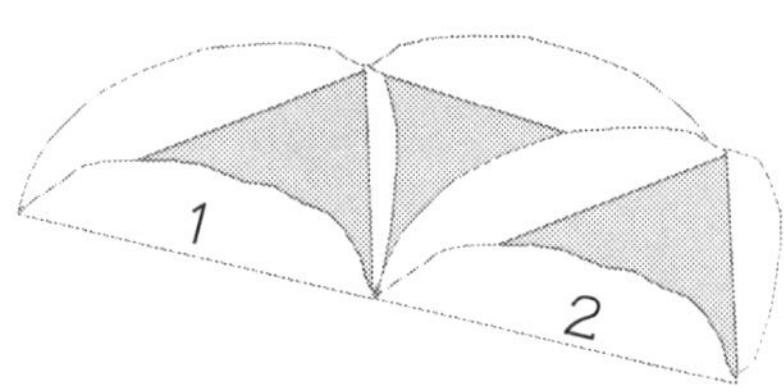

Flaps glued on the inside.

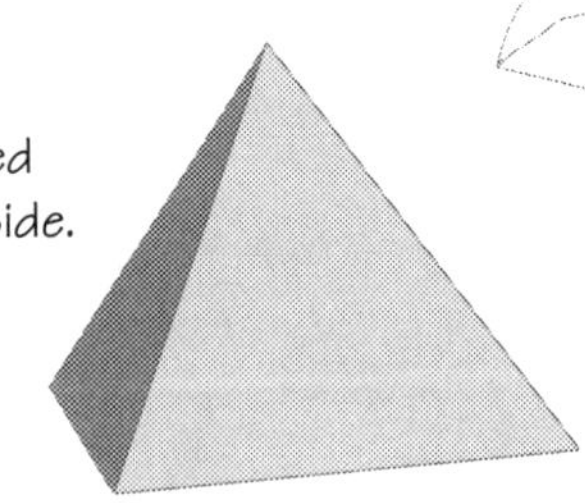

Flaps glued on the outside.

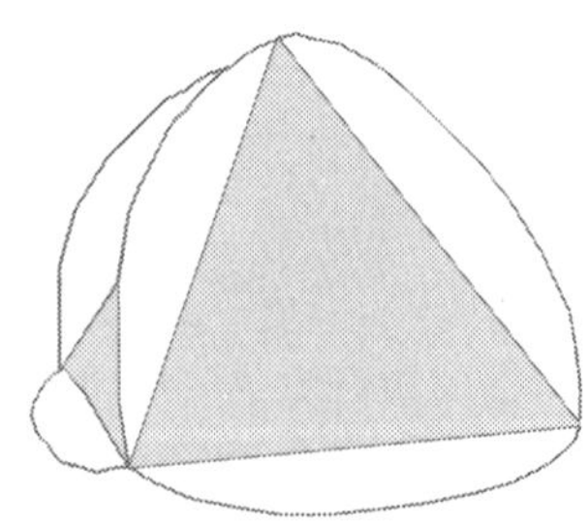

6. Glue the flaps of the fourth circle to the bottom of the figure. If the flaps have been glued to the outside, they may be trimmed with scissors.

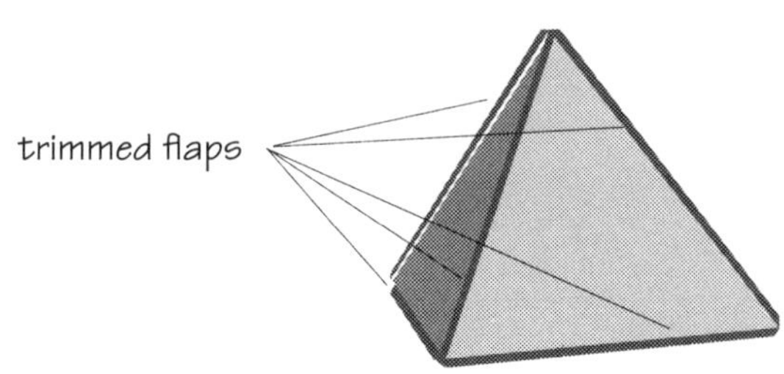

Pattern for Making the Tetrahedron, Octahedron, and Icosahedron

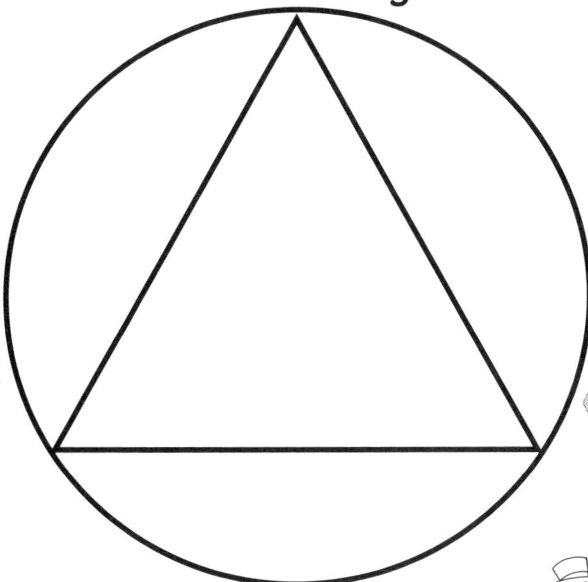

Tetrahedron
(4 triangles)

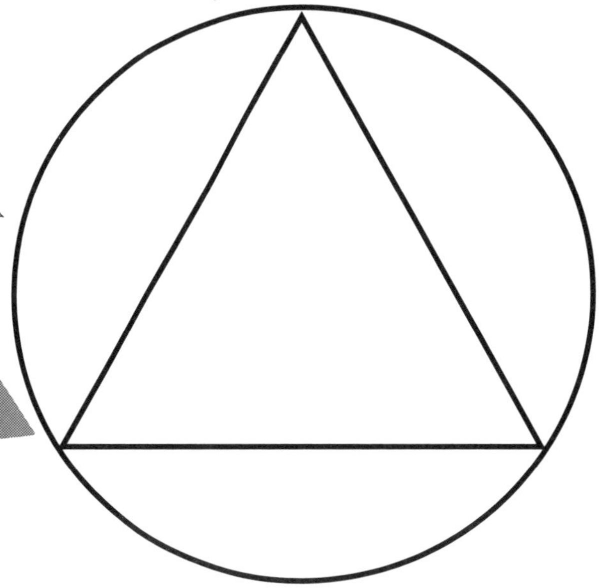

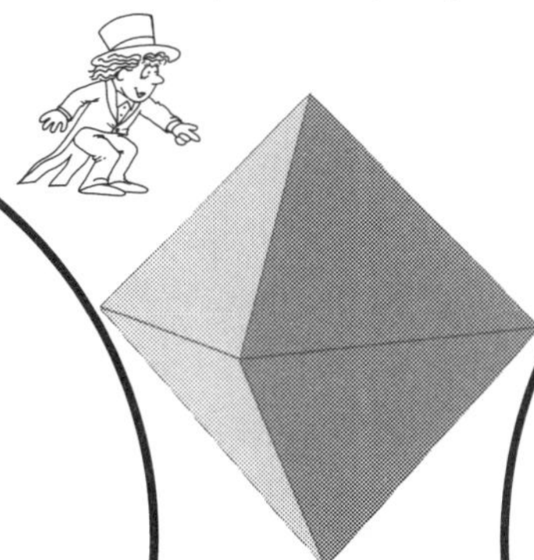

Octahedron
(8 triangles)

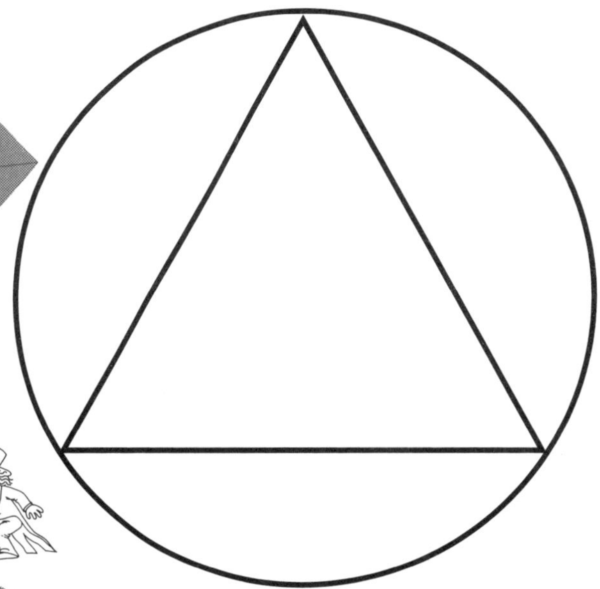

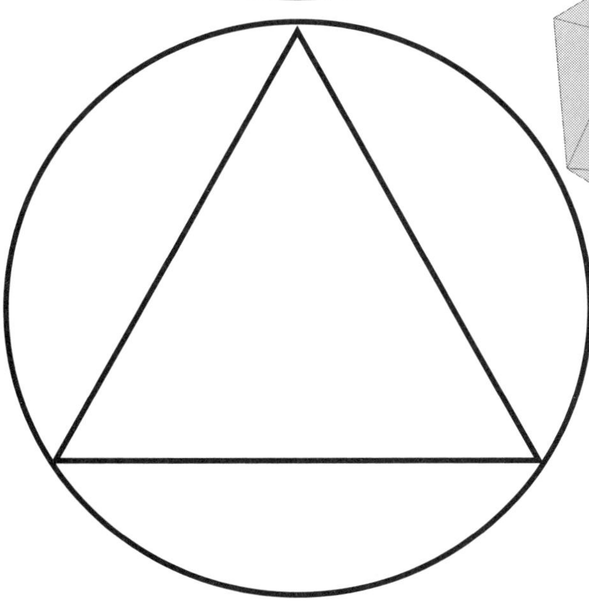

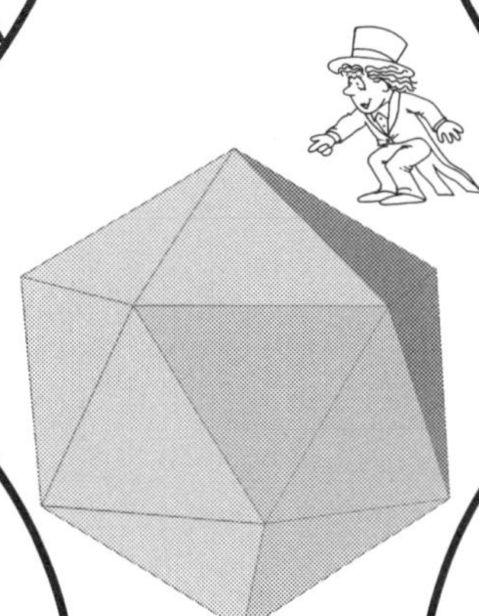

Icosahedron
(20 triangles)

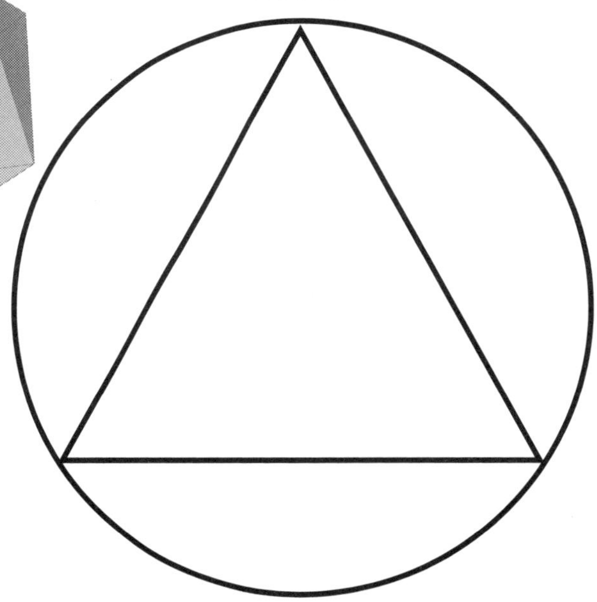

Pattern for Making the Hexahedron

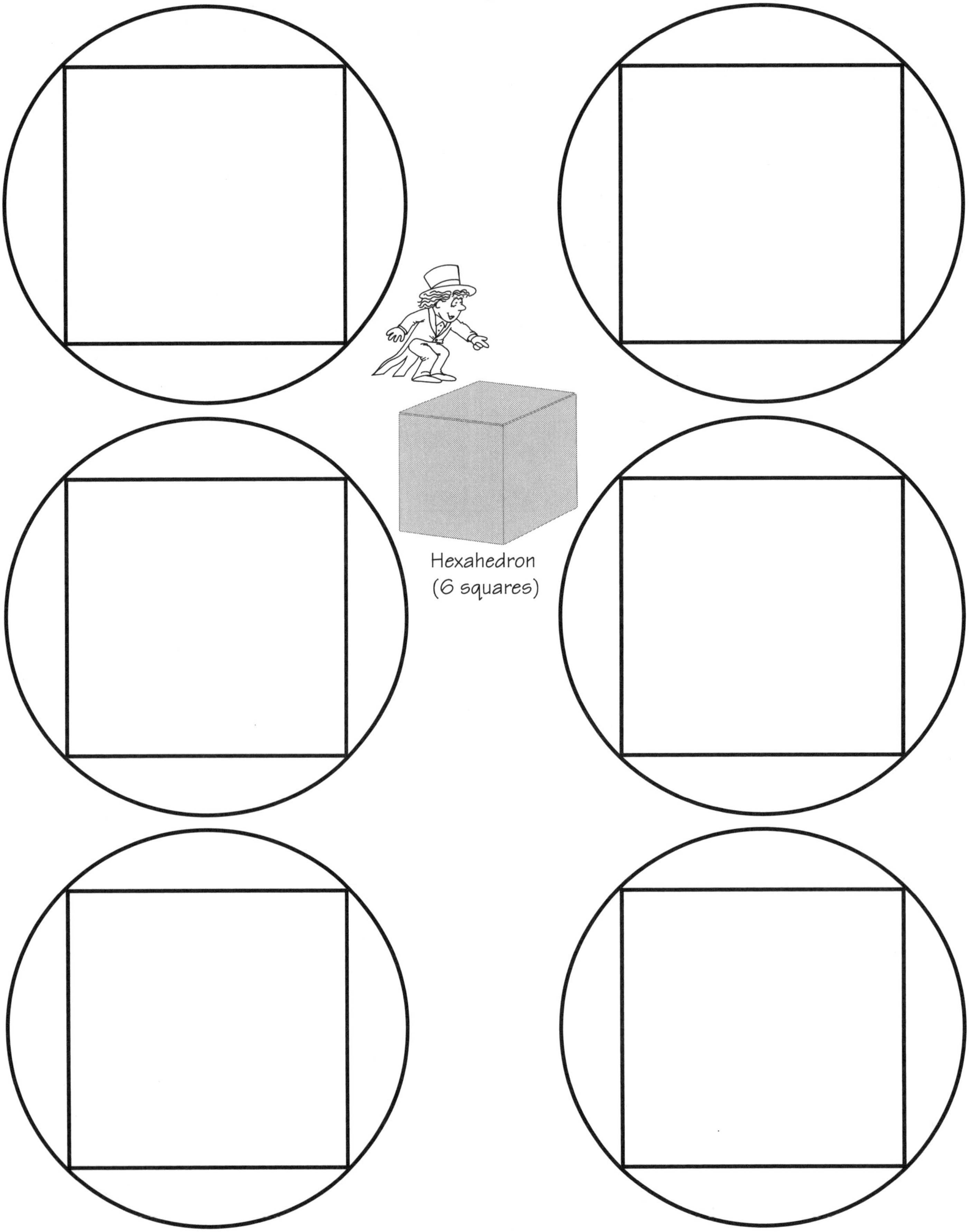

Hexahedron
(6 squares)

Pattern for Making the Dodecahedron

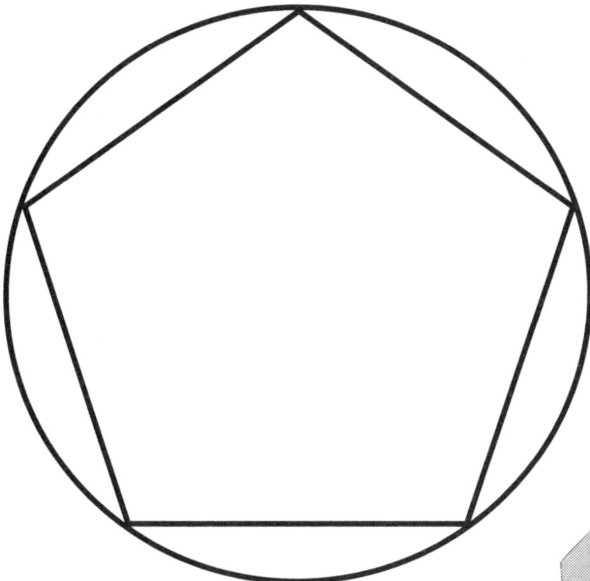

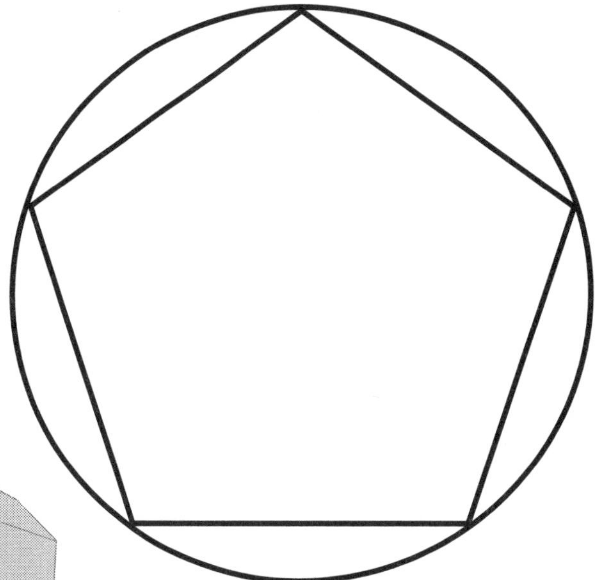

Dodecahedron
(12 pentagons)

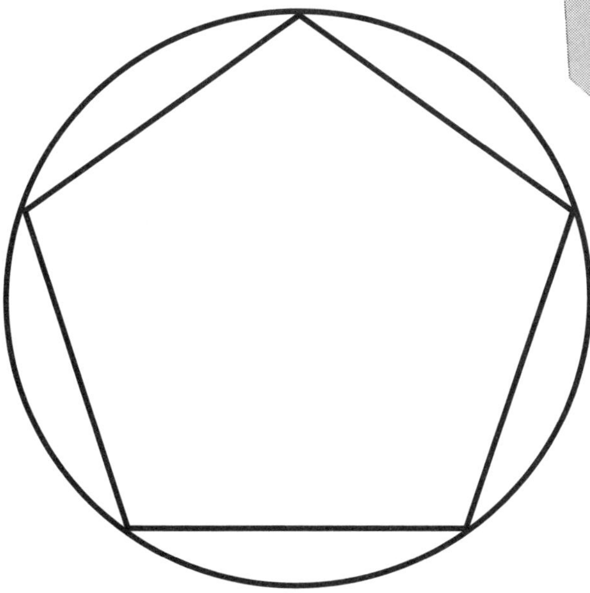

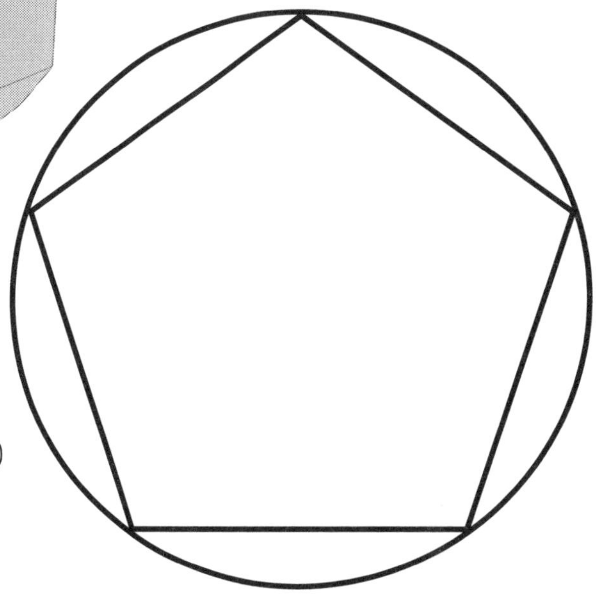

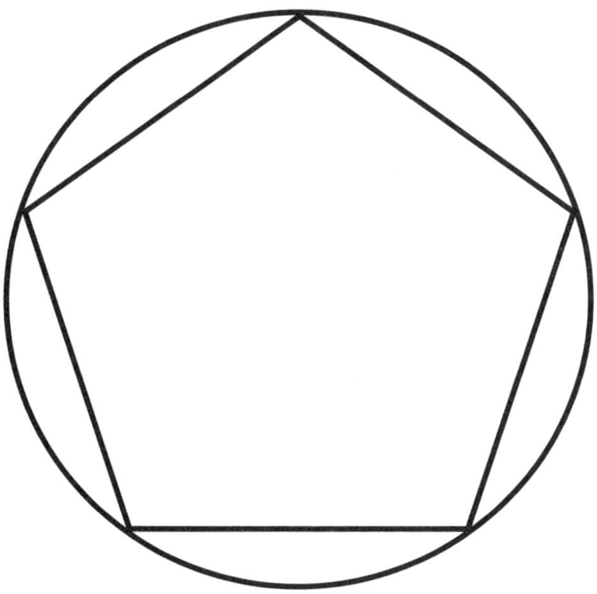

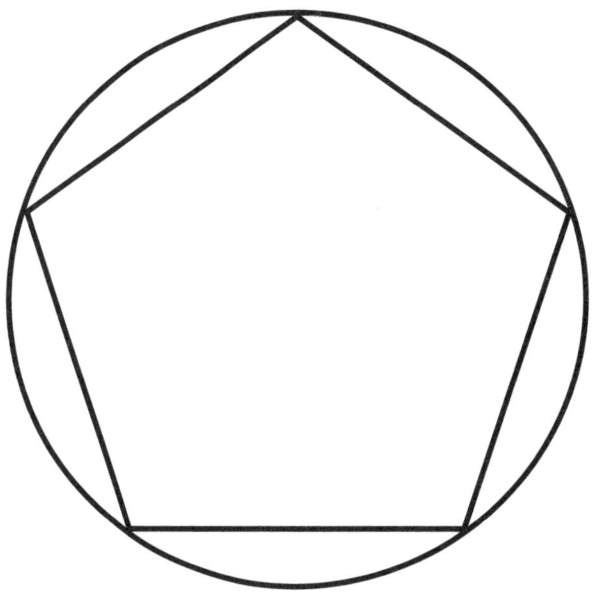

A Walk Along the Edges of the Platonic Solids In Which We Count the Number of Edges, Vertices, and Faces.

	Geometric Solid	Number of Faces (F)	Number of Vertices (V)	Number of Edges (E)	F + V	E + 2
	tetrahedron					
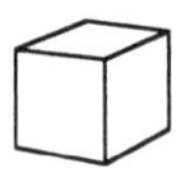	hexahedron					
	octahedron					
	dodecahedron					
	icosahedron					
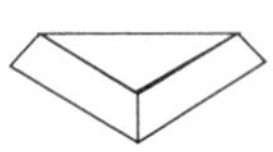	truncated tetrahedron					

Describe in your own words the relationship between the number of faces, vertices, and edges of the five Platonic solids and the truncated tetrahedron.

Write a formula for your description.

Apply the formula to a cone.  What did you find out?

Find a solid shape for which the formula isn't true?

Amazing Circle Geometry

Amazing Circle Geometry

What follows on these pages is a comparison between the folds made with The Amazing Circle and the corresponding geometry concepts. The Amazing Circle lesson number in which those concepts are addressed is located in the illustrated circle. For example, (4) indicates that the geometric concept is addressed in lesson number four.

Geometry	The Amazing Circle

Geometry

Plane

A geometric plane is thought of as a completely flat surface (no thickness) with no edges but extending on and on, without end, into space.

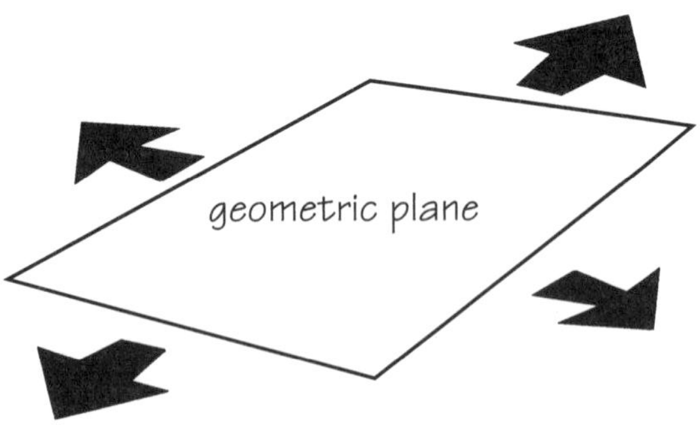

The portion of the plane enclosed by a circle is called a circular region.

The distance around a circle is called the circumference.

The Amazing Circle

Plane

The Amazing Circle is a circle cut from a model of a plane, a flat sheet of paper.

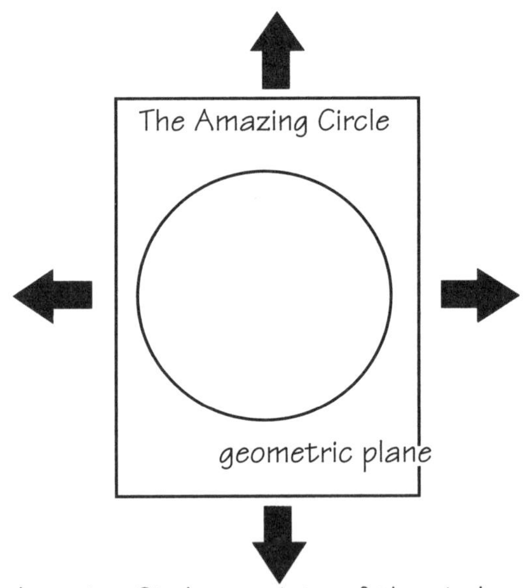

An Amazing Circle consists of the circle and the circular region enclosed by it.

The Amazing Circle = circle + circular region

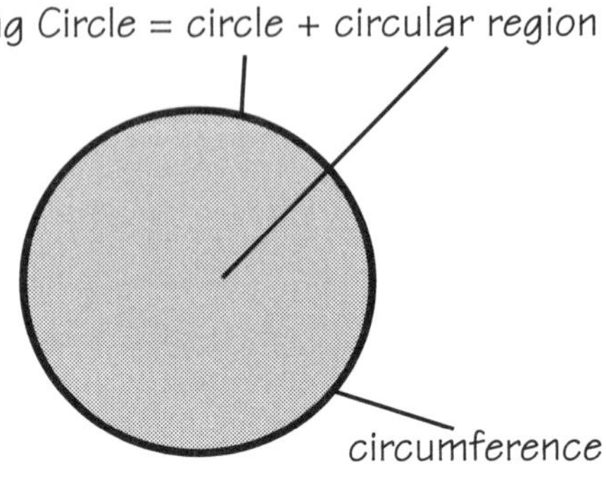

Amazing Circle Geometry

Geometry	The Amazing Circle

Point

A geometric point represents a single, exact location. A geometric point is represented by a small dot.

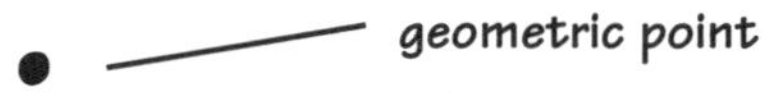

Uppercase letters are used to name points

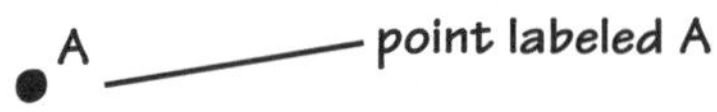

Point A is on the circle. Point B is outside the circle. Points C and D are inside the circle. Point C is a special point; it is the only point inside the circle that is the same distance from every point on the line that forms the circle. Point C is called the center of the circle.

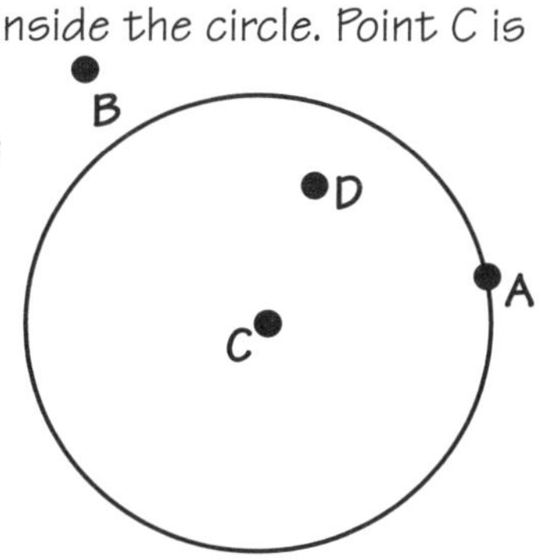

Line

A straight line can always be drawn through any two points.

The arrows mean that the line extends indefinitely in the direction indicated by the arrows.

$\overleftrightarrow{AB}$ is the symbol for the line passing through the points A and B.

Line Segment

That portion of the line between and including the endpoints A and B is a line segment

$\overline{AB}$ is the symbol for the line segment 

The line segment joining any two points on the circumference of a circle is called a chord.

Point

Points A and B are on the Amazing Circle and point C is the center of the Amazing Circle.

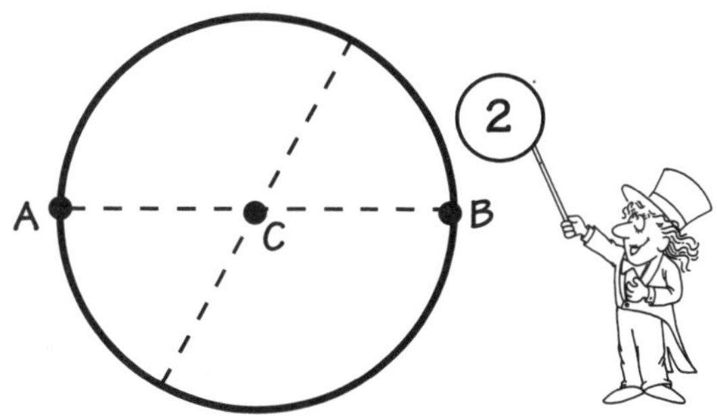

Line

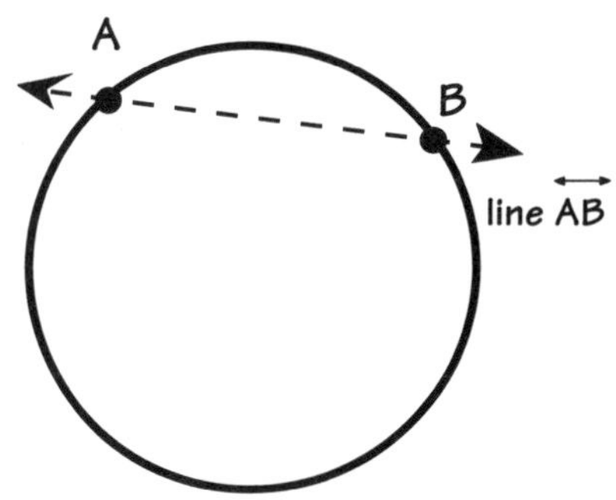

Line Segment

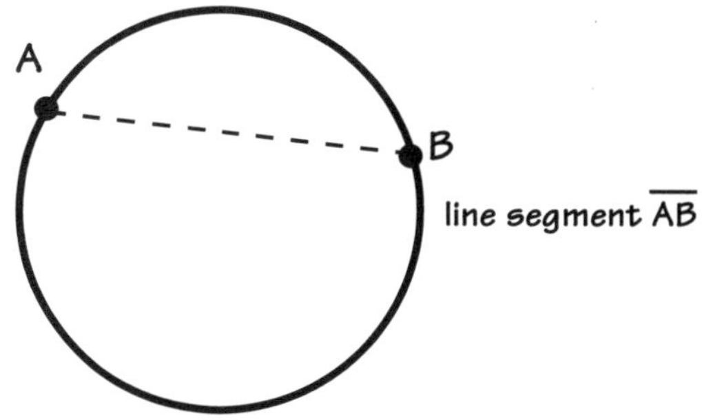

The line segment $\overline{AB}$ is a chord

Amazing Circle Geometry

Geometry

Diameter

The longest line segment, or chord, in a circle is called a diameter.

$\overline{AB}$ is a diameter of the circle.

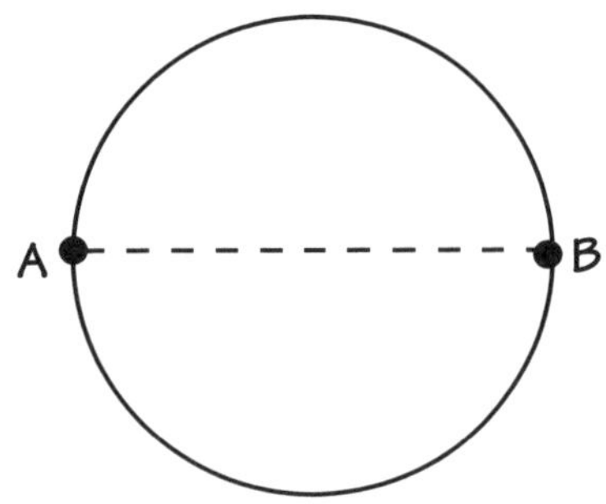

The diameter of a circle produces two semi-circles and two semi-circular regions.

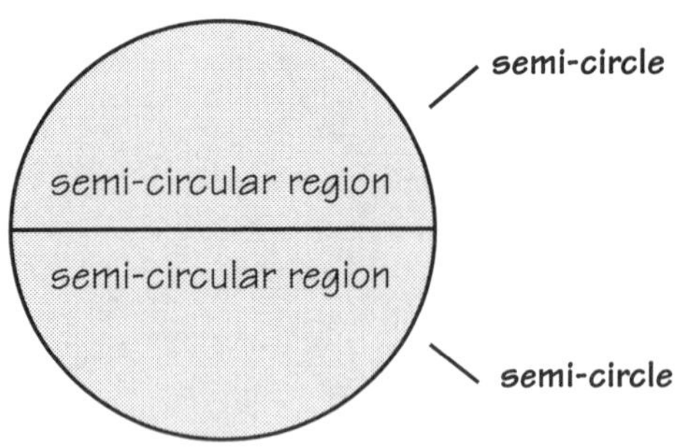

Intersecting Lines

Lines in the same plane either intersect or are parallel.

$\overleftrightarrow{AB}$ intersects $\overleftrightarrow{CD}$ at the point E

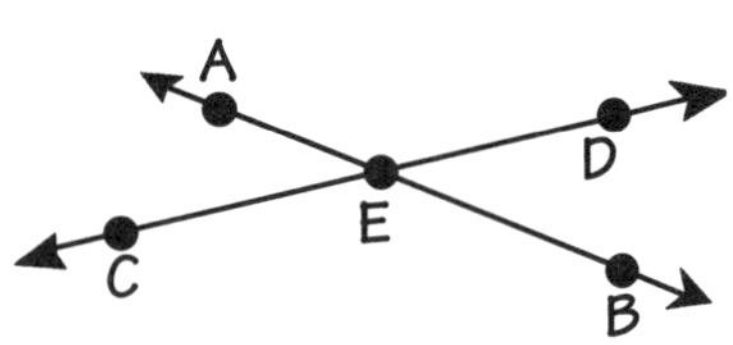

The Amazing Circle

Diameter

$\overline{AB}$ is a diameter of the circle.

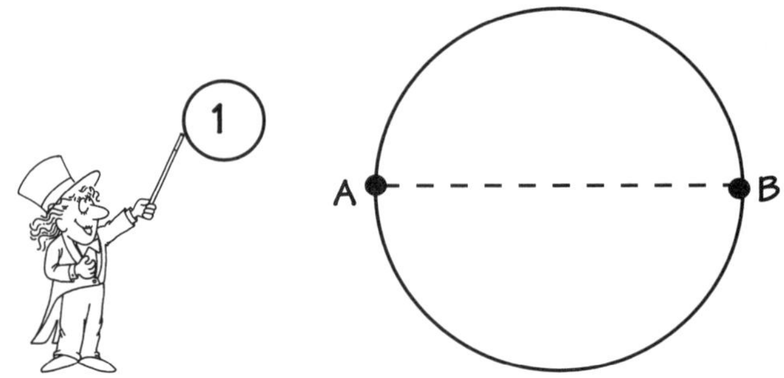

The circle has been folded along diameter $\overline{AB}$ to make a semi-circle.

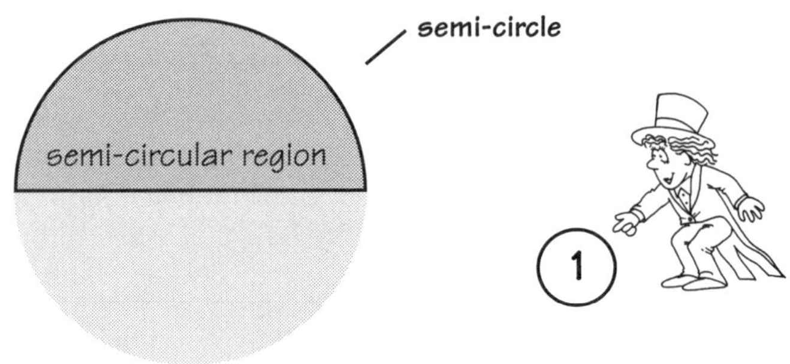

Intersecting Lines

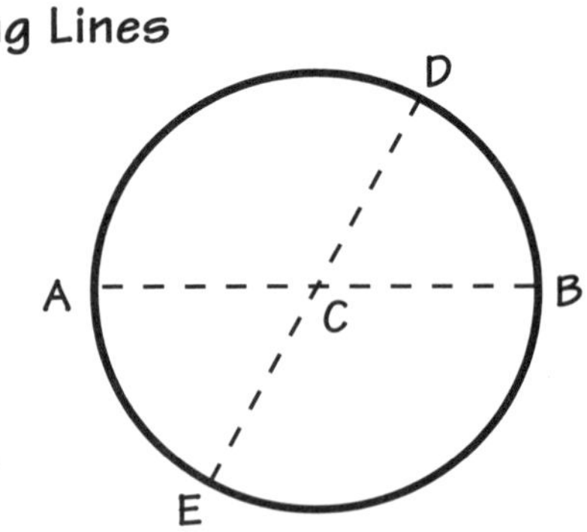

and

The diameter $\overline{AB}$ intersects the diameter $\overline{DE}$ at the center, C, of the circle.

Amazing Circle Geometry

Geometry	The Amazing Circle

Congruence

Geometric figures that have the same size and the same shape are called congruent figures.

The symbol ≅ means is congruent to. The symbol ≇ means is not congruent to.

Any two line segments are congruent if they have the same length.

$\overline{AB} \cong \overline{CD}$

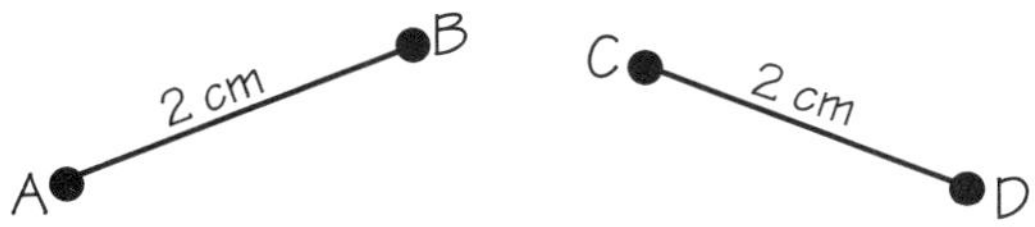

Any two angles are congruent if they have the same measure.

$\angle A \cong \angle B$

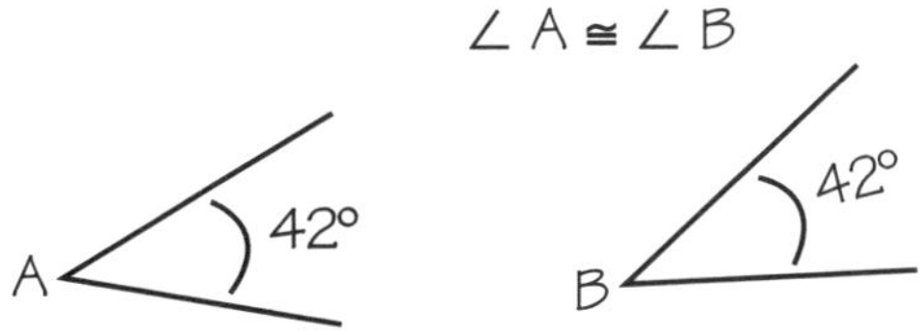

Any two circles are congruent if their radii are congruent.

$\overline{CA}$ and $\overline{CB}$ are radii of two different circles.

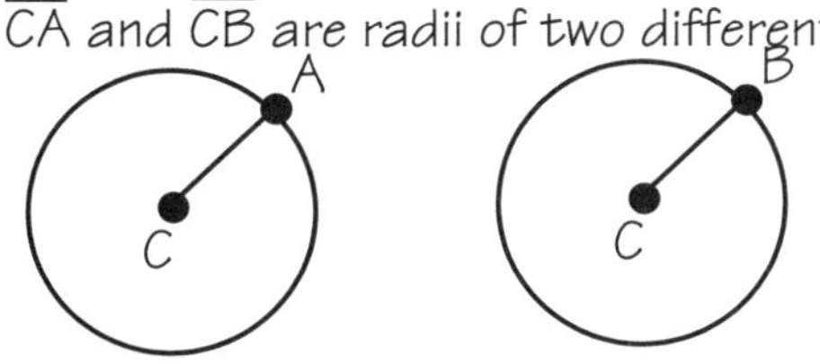

$\overline{CA} \cong \overline{CB}$, therefore the two circles are congruent.

Any two polygons are congruent if their corresponding sides and their corresponding angles are congruent. Corresponding sides and angles are in the same relative position.

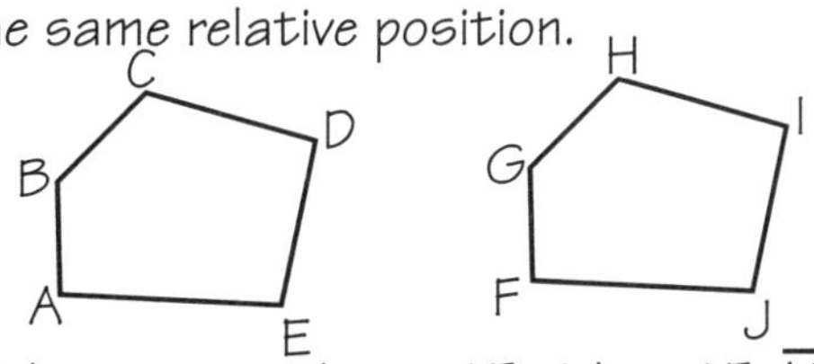

$\angle A$ corresponds to $\angle F$, $\angle A \cong \angle F$, $\overline{AB} \cong \overline{FG}$
$\angle B$ corresponds to $\angle G$, $\angle B \cong \angle G$, $\overline{BC} \cong \overline{GH}$
$\angle C$ corresponds to $\angle H$, $\angle C \cong \angle H$, $\overline{CD} \cong \overline{HI}$
$\angle D$ corresponds to $\angle I$, $\angle D \cong \angle I$, $\overline{DE} \cong \overline{IJ}$
$\angle E$ corresponds to $\angle J$, $\angle E \cong \angle J$, $\overline{EA} \cong \overline{JF}$

Congruence

Every Amazing Circle activity incorporates congruent figures. These are just a few examples of congruent line segments, congruent angles, and congruent figures contained in the Amazing Circle.

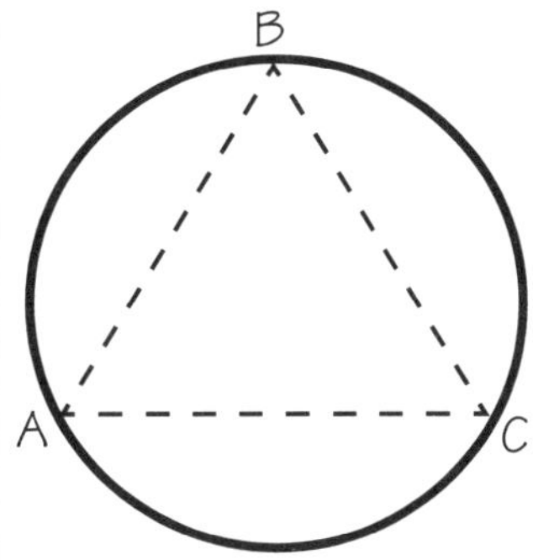

$\angle A \cong \angle B \cong \angle C$
$\overline{AB} \cong \overline{BC} \cong \overline{CA}$

sector 1 ≅ sector 3
sector 2 ≅ sector 4

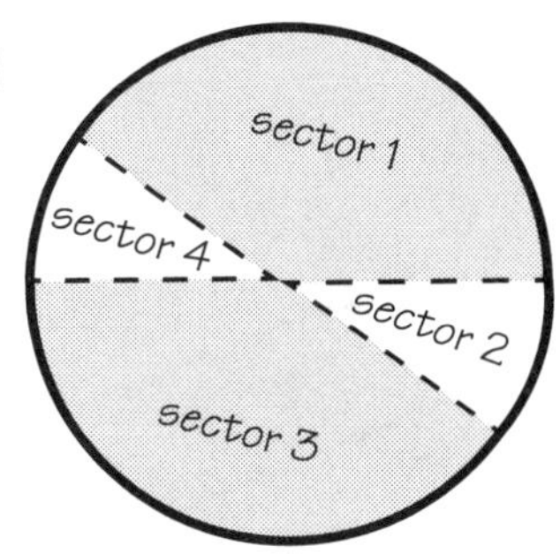

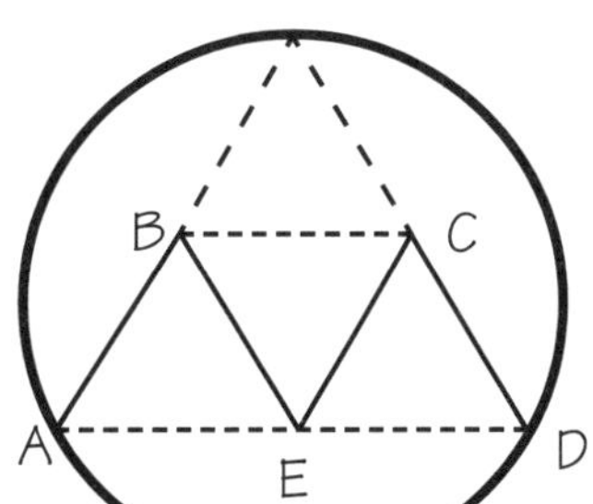

$\Delta ABE \cong \Delta BCE \cong \Delta CDE$

Amazing Circle Geometry

Geometry	The Amazing Circle

Geometry

Perpendicular Lines

$\overleftrightarrow{AB}$ is perpendicular to $\overleftrightarrow{ED}$ at the point C.

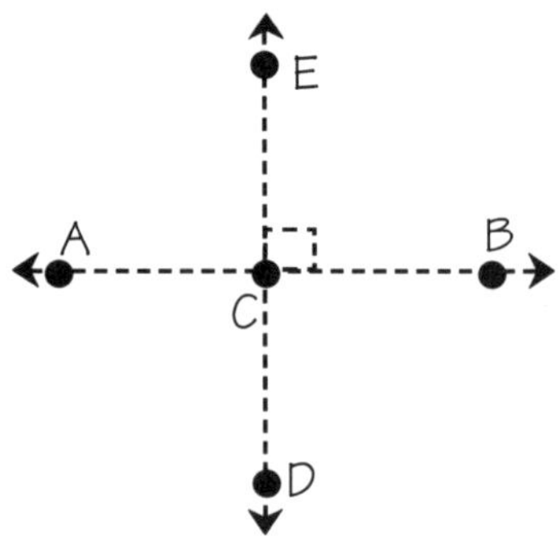

Perpendicular lines intersect at right angles.

A right angle equals 90°.

The square corner symbol at the intersection, point C in the diagram, denotes a right angle.

The symbol for perpendicular lines is ⊥.

Parallel Lines

Lines on a flat surface (in the same plane) that never intersect (have a point in common) are called parallel lines.

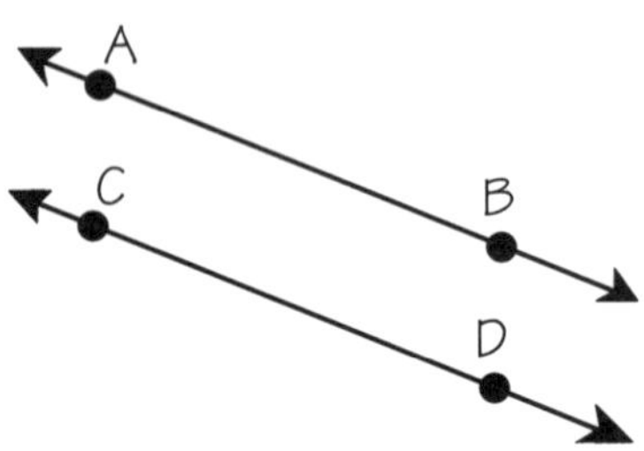

$\overleftrightarrow{AB}$ is parallel to $\overleftrightarrow{CD}$.

The symbol for parallel lines is ll

The Amazing Circle

Perpendicular Lines

$\overline{AB}$ is perpendicular to $\overline{ED}$ at the point C.

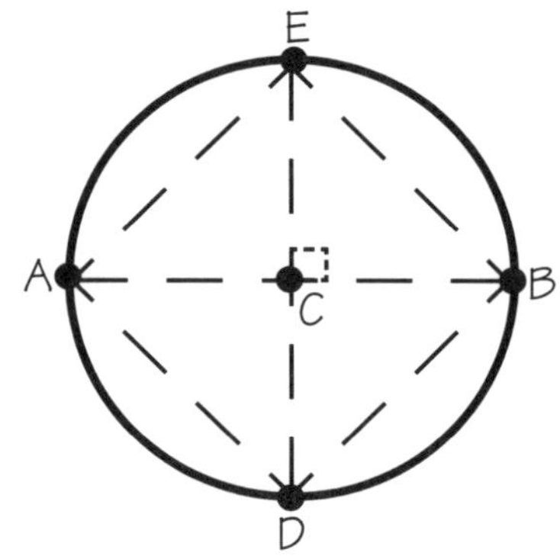

$\overline{AB} \perp \overline{ED}$

Parallel Lines

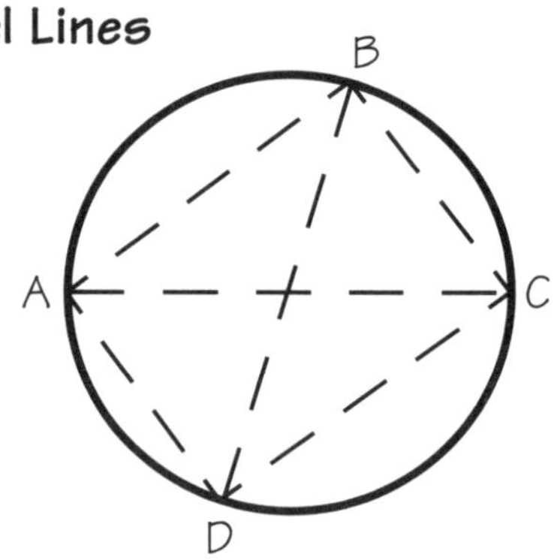

$\overline{AB}$ is parallel to $\overline{DC}$.
$\overline{AD}$ is parallel to $\overline{BC}$.

$\overline{AB}$ ll $\overline{DC}$
$\overline{AD}$ ll $\overline{BC}$

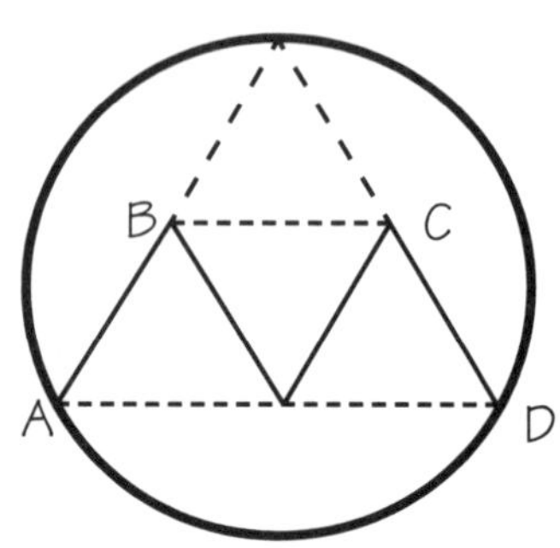

$\overline{BC}$ is parallel to $\overline{AD}$.

$\overline{BC}$ ll $\overline{AD}$

Amazing Circle Geometry

Geometry

Parallel Lines (continued)

The Amazing Circle

Parallel Lines (continued)

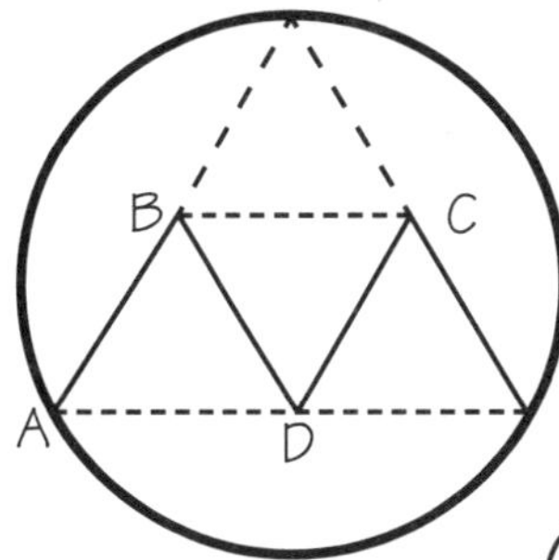

$\overline{AD}$ is parallel to $\overline{BC}$.
$\overline{AB}$ is parallel to $\overline{DC}$.

$\overline{AD} \parallel \overline{BC}$
$\overline{AB} \parallel \overline{DC}$

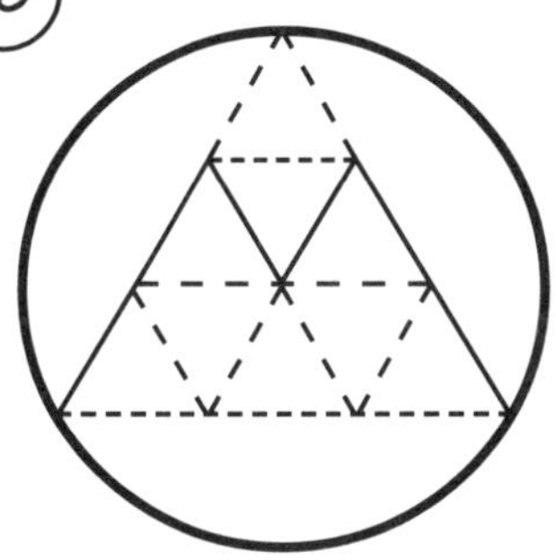

$\overline{AD}$ is parallel to $\overline{BC}$.

$\overline{AD} \parallel \overline{BC}$

Amazing Circle Geometry

Geometry

Radius

One-half a diameter is called a radius. $\overline{CA}$, $\overline{CB}$, $\overline{CD}$, and $\overline{CE}$ are radii.

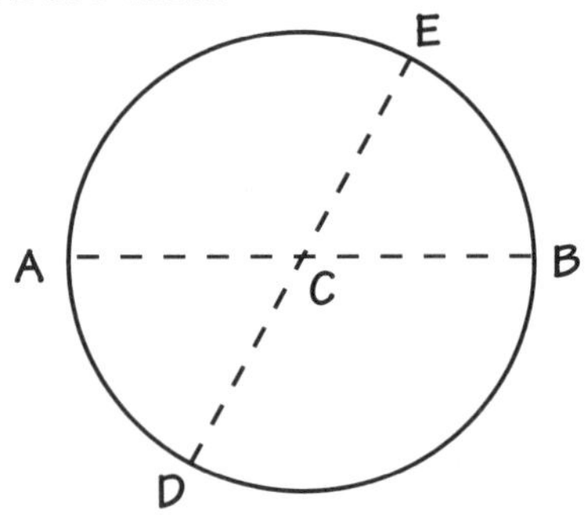

All radii of the same circle are congruent.

Arc

The two diameters, $\overline{AB}$ and $\overline{DE}$, cut the circle into four parts called arcs.

The symbol for arc is $\frown$.

$\overset{\frown}{AE}$, $\overset{\frown}{EB}$, $\overset{\frown}{BD}$, and $\overset{\frown}{DA}$ are arcs of the circle.

Arcs greater than a semi-circle are called major arcs and arcs less than a semi-circle are called minor arcs.

$\overset{\frown}{AEBD}$ is a major arc and $\overset{\frown}{DA}$ is a minor arc of the circle.

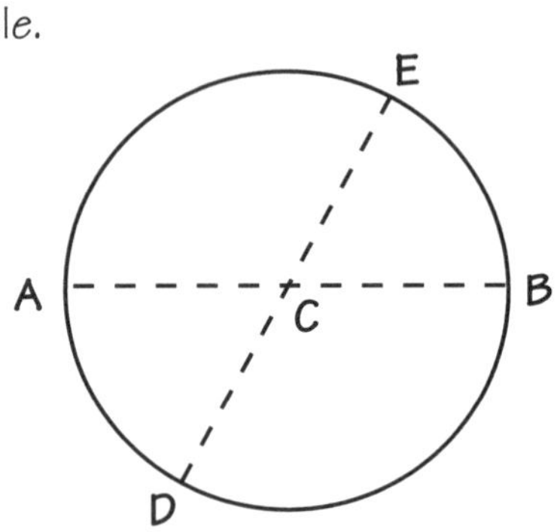

Other major and minor arcs can also be named.

The Amazing Circle

Radius

$\overline{CA}$, $\overline{CE}$, $\overline{CB}$ and $\overline{CD}$ are radii.
$\overline{CA} \cong \overline{CE} \cong \overline{CB} \cong \overline{CD}$

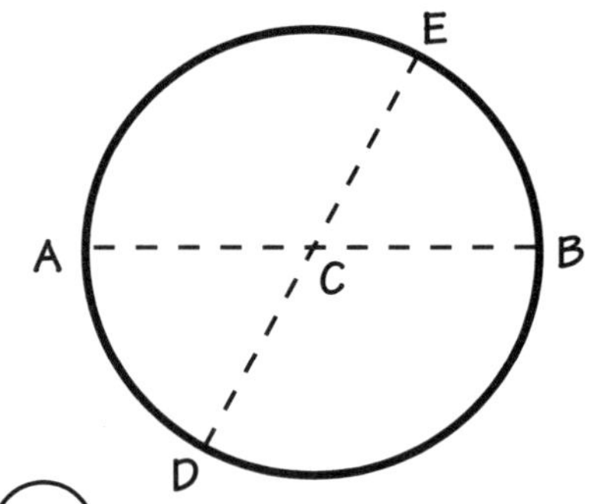

Arc

$\overset{\frown}{AE}$, $\overset{\frown}{EB}$, $\overset{\frown}{BD}$, and $\overset{\frown}{DA}$ are arcs of the circle

$\overset{\frown}{AEBD}$ is a major arc and $\overset{\frown}{AD}$ is a minor arc of the circle.

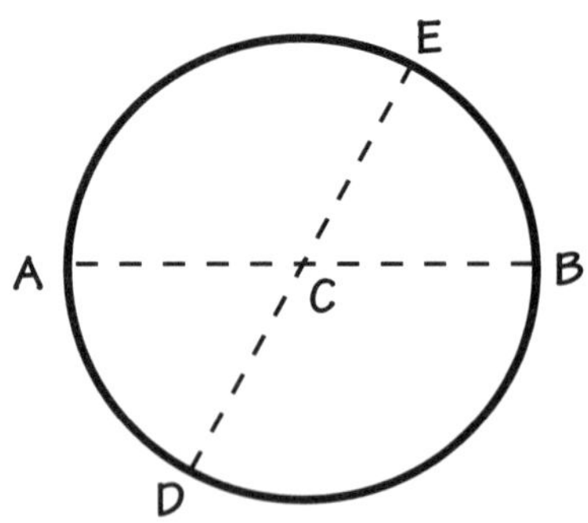

Other major and minor arcs can also be named.

Amazing Circle Geometry

Geometry

Sector

Two diameters create four sectors in a circle.

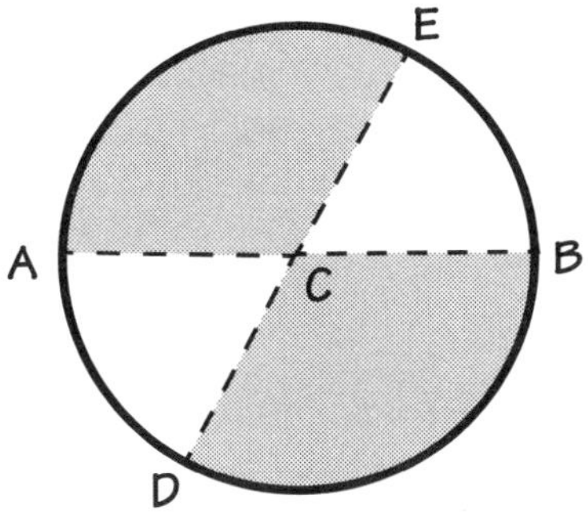

A sector is a region of a circle bounded by two radii and the intercepted arc.

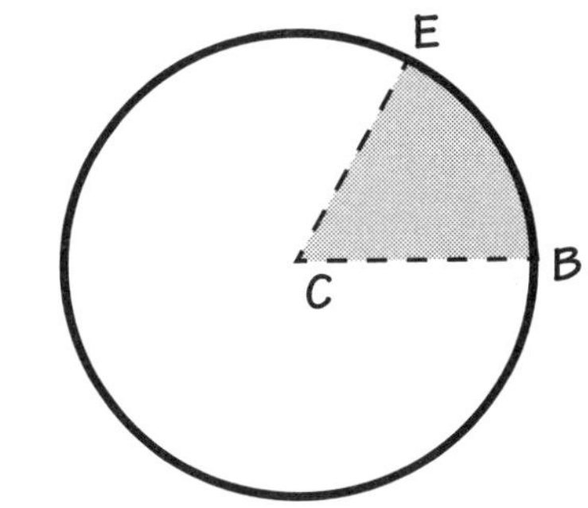

The radii, $\overline{CE}$ and $\overline{CB}$, and the intercepted $\overset{\frown}{\text{arc}}$ EB, form a sector of the circle.

Ray

A ray is that portion of a line starting at a point and going on forever in the direction indicated by the arrow.

$\overrightarrow{AB}$ is the symbol for the ray starting at point A, passing through point B, and continuing on forever in the direction indicated by the arrow.

The Amazing Circle

Sector

The two diameters, $\overline{AB}$ and $\overline{DE}$, create four sectors in the Amazing Circle.

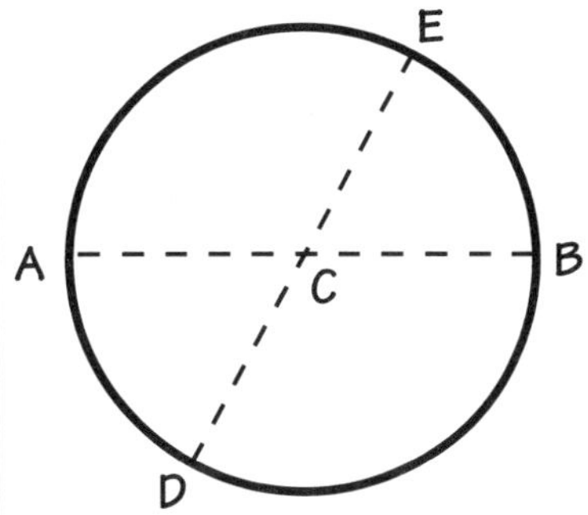

The radii, $\overline{CE}$ and $\overline{CB}$, and the intercepted $\overset{\frown}{\text{arc}}$ EB, form a sector of the circle.

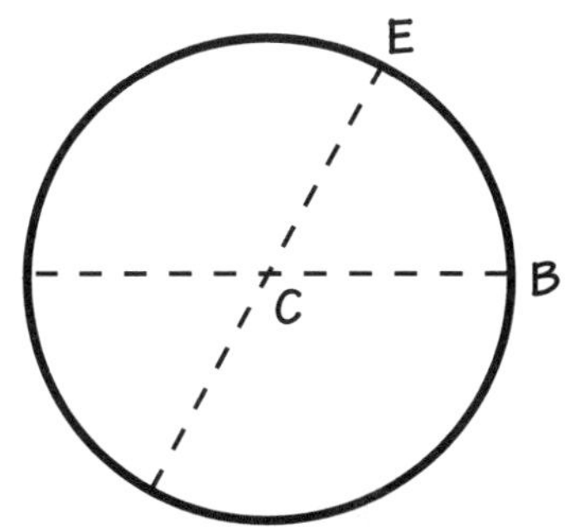

Ray

The Amazing Circle can only display part of a ray.

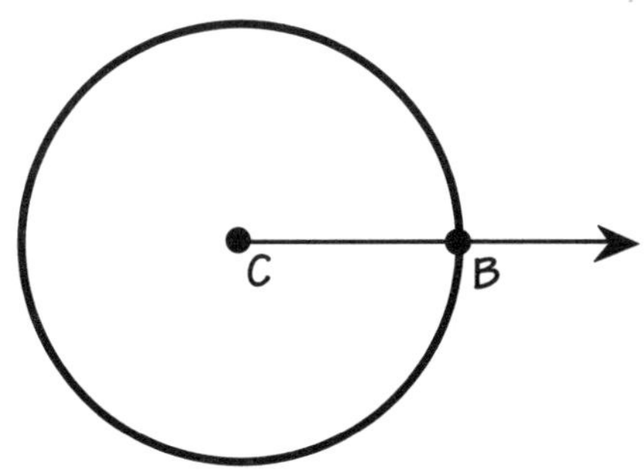

$\overrightarrow{CB}$ is a ray.
$\overline{CB}$ is the only portion of the ray the Amazing Circle contains.

Amazing Circle Geometry

Geometry

Angles

The figure formed by two different rays starting from the same endpoint is called an angle. The symbol for angle is ∠.

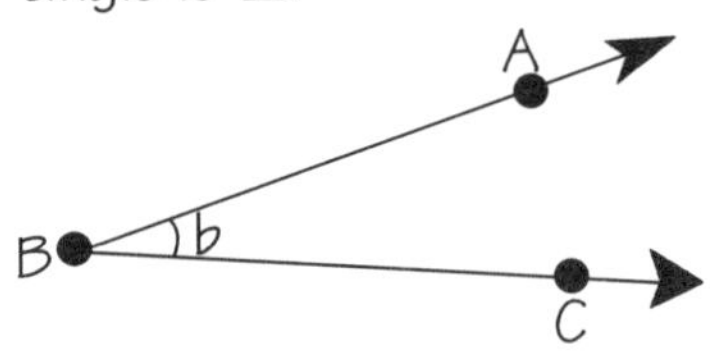

$\overrightarrow{BA}$ is one side of the angle.

$\overrightarrow{BC}$ is the other side of the angle.

The point B is the vertex of the angle.

The angle can be named in three ways:

1. By its sides and vertex, ∠ABC
2. By its vertex, ∠B
3. By the angle itself, ∠b

An angle with a measure less than 90° is an acute angle.

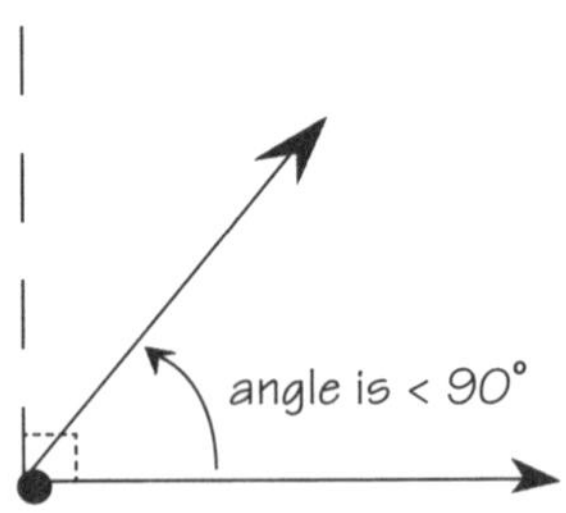

An angle with a measure greater than 90° but less than 180° is an obtuse angle.

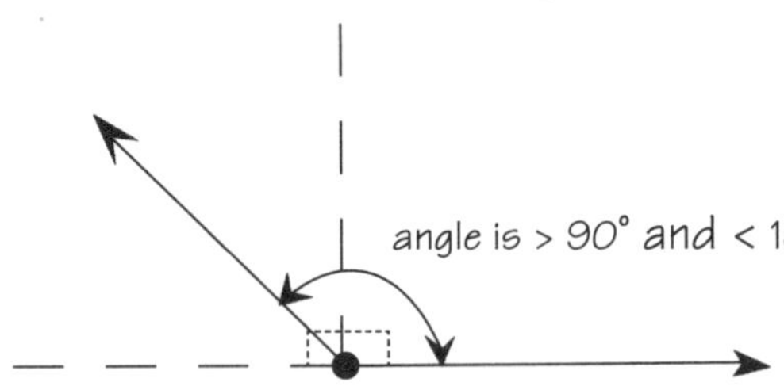

An angle with a measure equal to 180° is a straight angle.

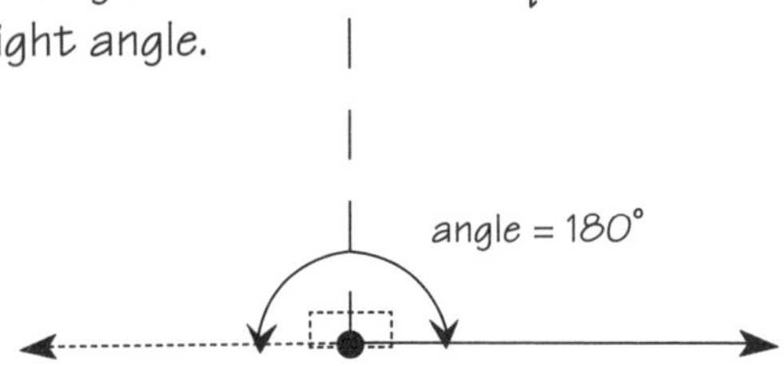

The Amazing Circle

Angles

In the Amazing Circle, angles are formed by line segments that can be thought of as portions of rays.

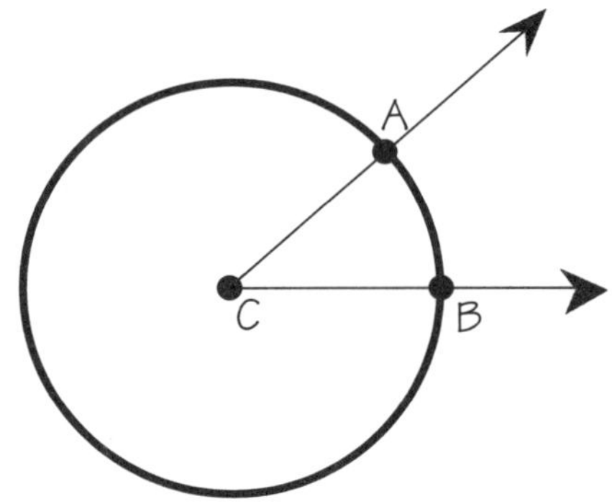

∠ACB is formed by the vertex C and the line segments, both portions of rays, $\overline{CA}$ and $\overline{CB}$.

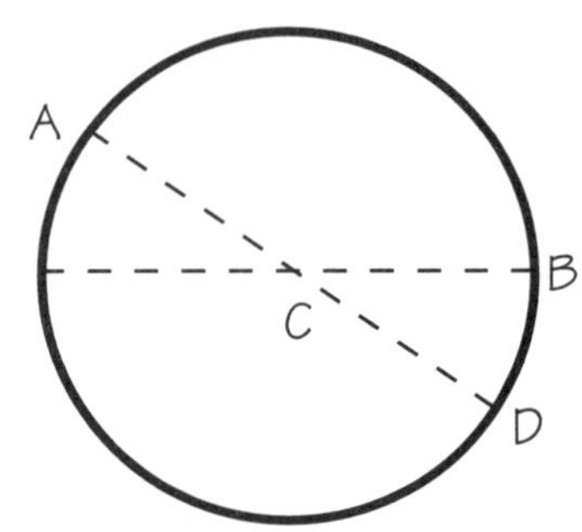

∠ACB + ∠DCB = 180° = straight angle

Amazing Circle Geometry

Geometry

Angles (continued)

Opposite angles

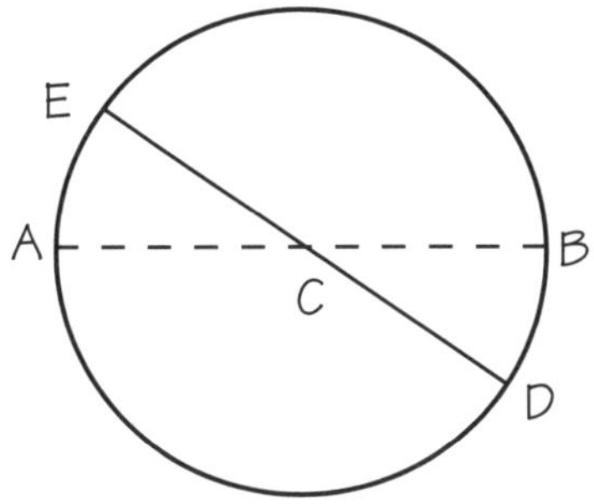

∠ ACE, is opposite ∠ BCD. Opposite angles are congruent. ∠ ACE ≅ ∠ BCD.

Central angle

The angle formed by two radii is called a central angle.

∠ ACE, ∠ ECB, ∠ BCD, ∠ DCA (and other overlapping angles) are central angles.

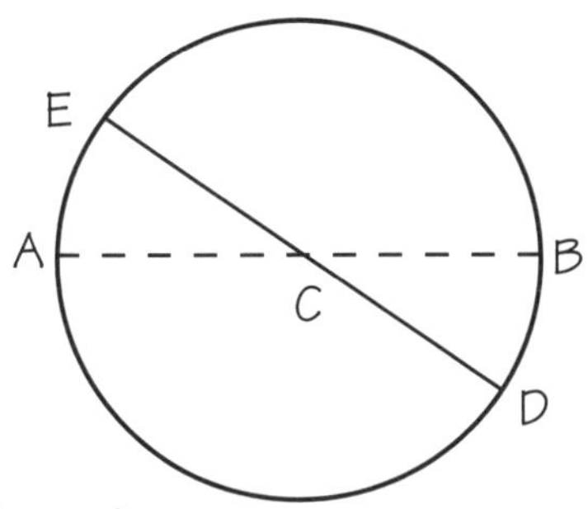

Inscribed angle

An angle whose vertex is on the circle and whose sides are chords is called an inscribed angle.

∠ ABC is an inscribed angle

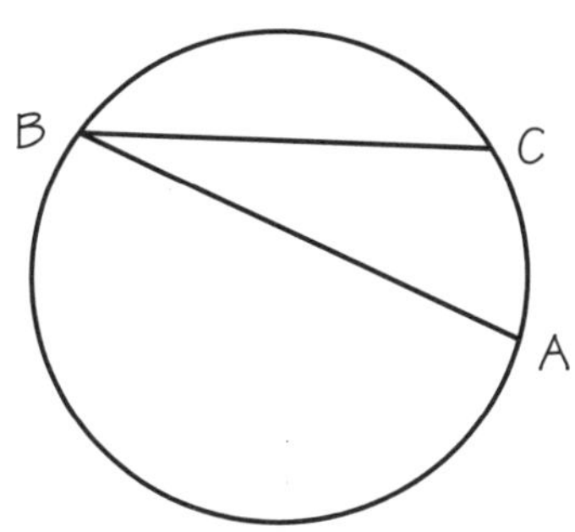

The Amazing Circle

Angles (continued)

Opposite angles

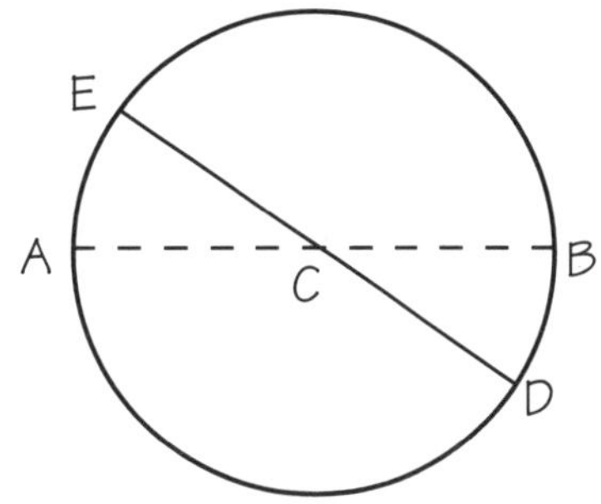

∠ ACE is opposite ∠ BCD. Opposite angles are congruent. ∠ ACE ≅ ∠ BCD.

Central angle

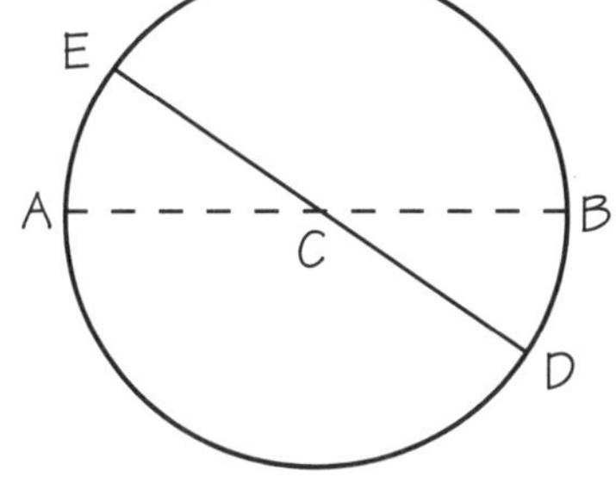

∠ ACE, ∠ ECB, ∠ BCD, ∠ DCA (and other overlapping angles) are central angles.

Inscribed angle

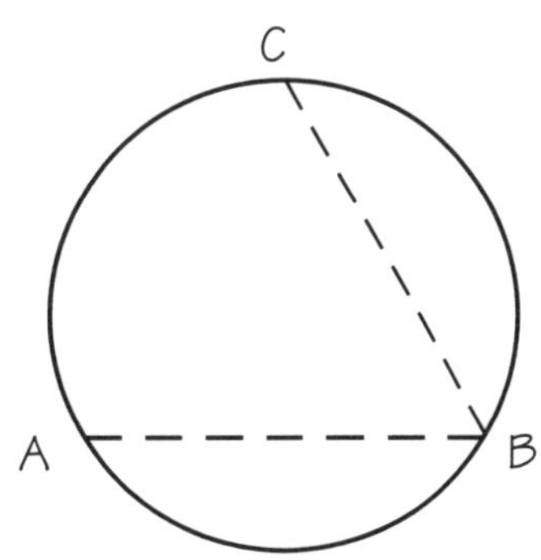

∠ ABC is an inscribed angle

Amazing Circle Geometry

Geometry

Polygon

A closed plane figure formed by the line segments that join — without crossing — more than two points not on the same straight line is called a polygon.

The pentagon, ⬠ ABCDE, is a polygon.

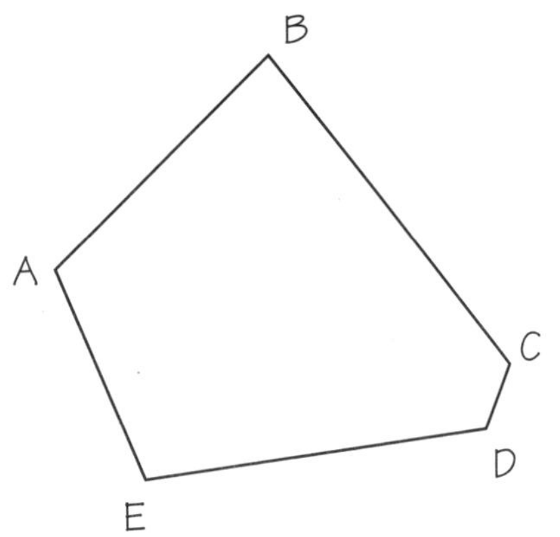

Diagonal

A line segment connecting any two vertices of a polygon not already connected is called a diagonal.

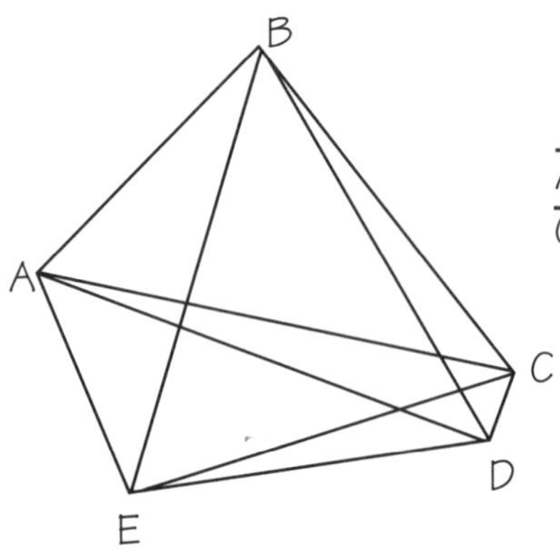

$\overline{AC}$, $\overline{AD}$, $\overline{BE}$, $\overline{BD}$, and $\overline{CE}$ are diagonals

Polygon Classification

Polygons classified according to the number of sides

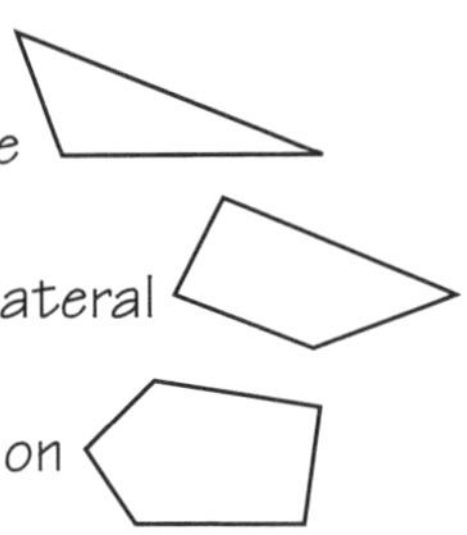

3 sides — triangle

4 sides — quadrilateral

5 sides — pentagon

The Amazing Circle

Polygon

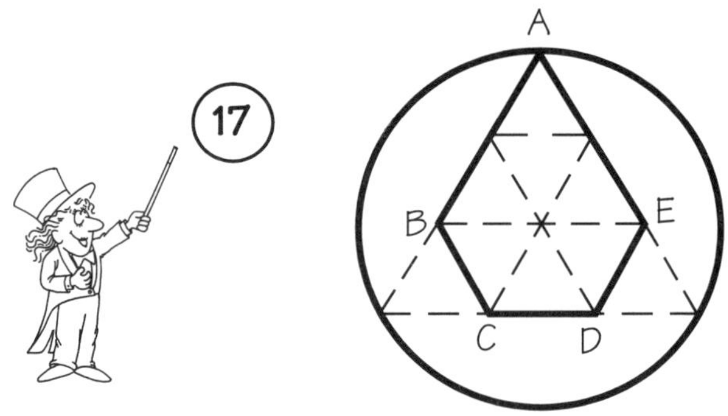

The polygon ABCDE is a pentagon.

Diagonal

$\overline{AC}$ and $\overline{BD}$ are diagonals of the rectangle ▭ ABCD.

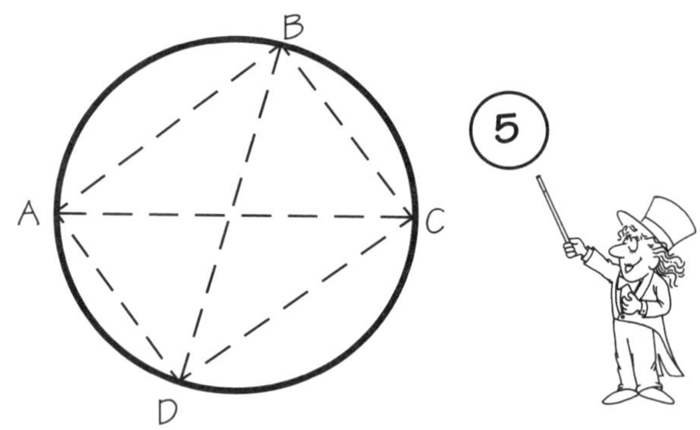

Polygon Classification

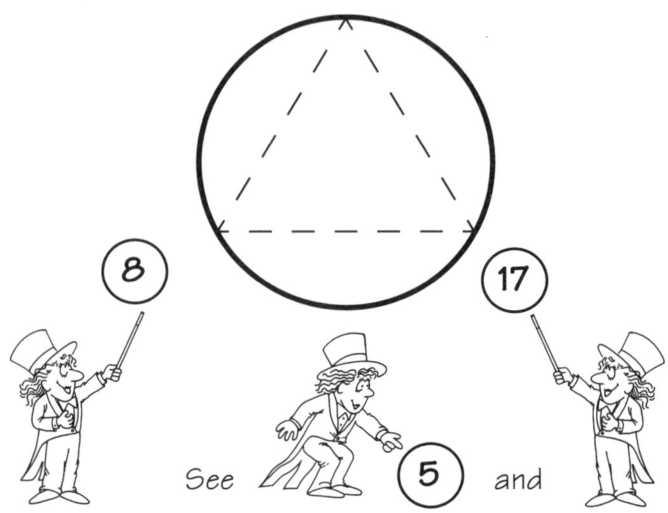

Amazing Circle Geometry

Geometry

Polygon Classification (continued)

6 sides — hexagon

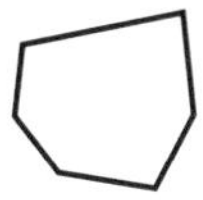

7 sides — heptagon

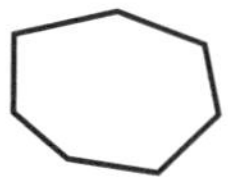

8 sides — octagon

9 sides — nonagon

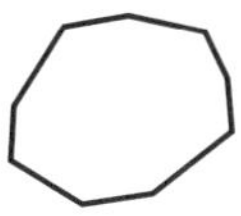

10 sides — decagon

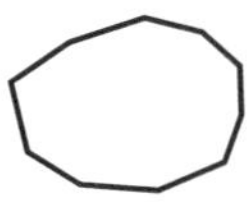

A polygon in which all of the angles are congruent is called equiangular.

Every angle in the pentagon has a measure of 108°

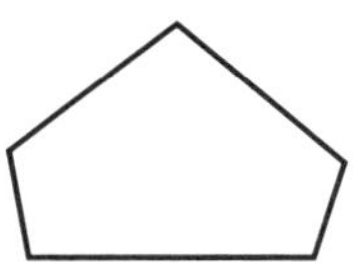

A polygon in which all of the sides are congruent is called equilateral.

Every side in the hexagon has a measure of 1 cm.

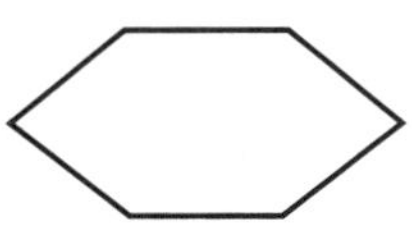

A polygon that is both equiangular and equilateral is called a regular polygon.

A regular hexagon with 6 congruent angles and 6 congruent sides.

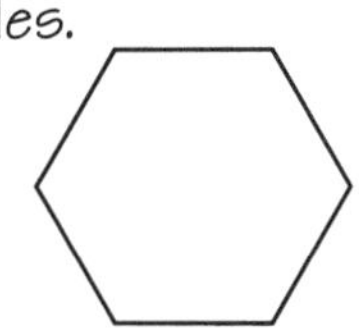

The Amazing Circle

Polygon Classification (continued)

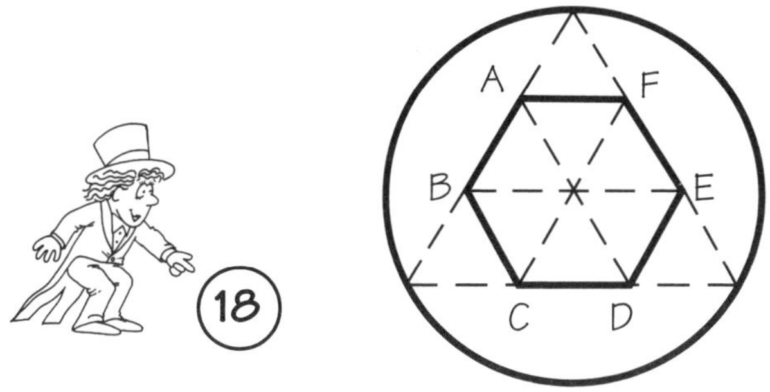

The polygon ABCDEF is a hexagon.

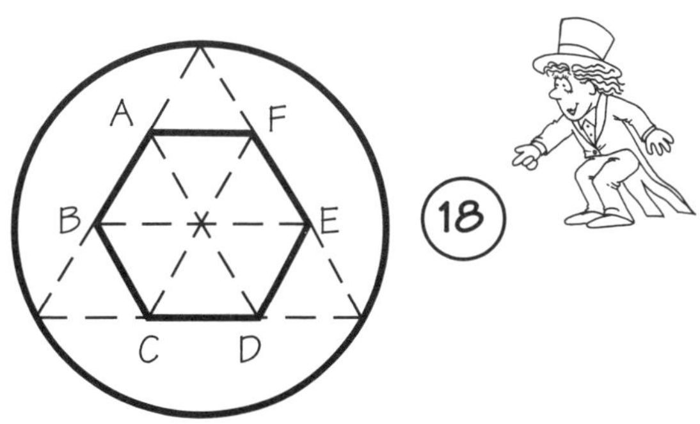

The polygon ABCDEF is a regular hexagon.

Amazing Circle Geometry

Geometry

Triangle Classification by Sides

A scalene triangle is a triangle having no congruent sides.

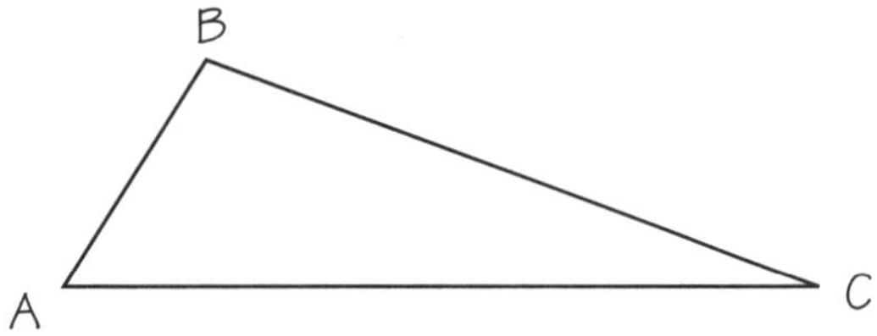

In Δ ABC, none of the sides $\overline{AB}$, $\overline{BC}$, or $\overline{CA}$ are congruent.

An isosceles triangle is a triangle having two congruent sides.

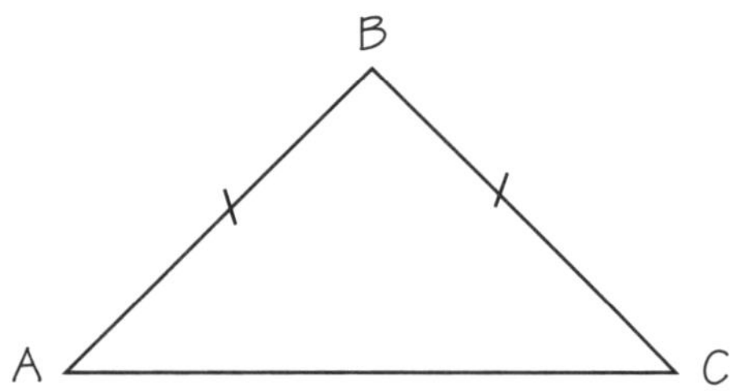

In Δ ABC, the sides $\overline{AB}$ and $\overline{BC}$ are congruent. The congruent sides are denoted on the figure with a slash (/) mark.

An equilateral triangle is a triangle having three congruent sides.

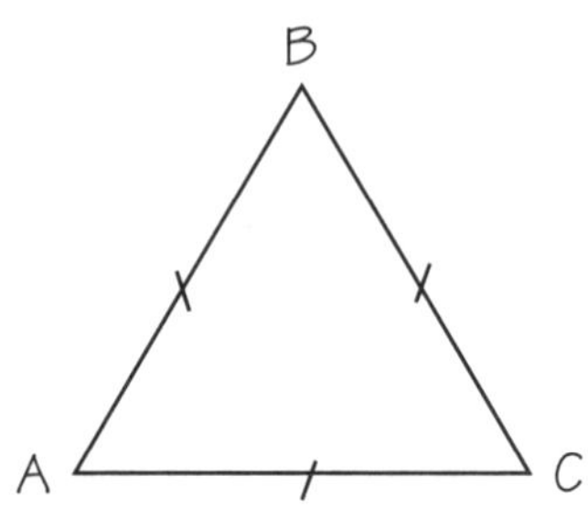

In Δ ABC, the sides $\overline{AB}$ and $\overline{BC}$, and $\overline{CA}$ are congruent.

The Amazing Circle

Triangle Classification by Sides

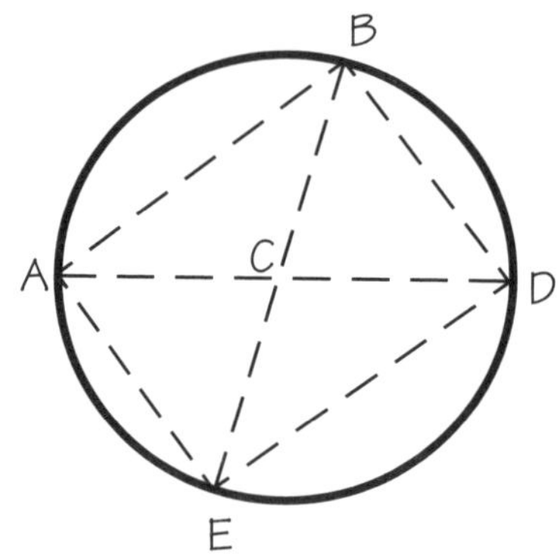

5

Δ ACB is an isosceles triangle. $\overline{AC} \cong \overline{CB}$.

7

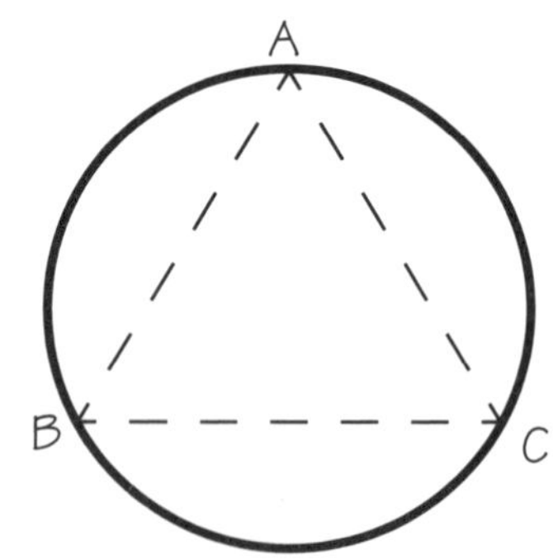

In Δ ABC, the sides $\overline{AB}$ and $\overline{BC}$ and $\overline{CA}$ are congruent. Δ ABC is an equilateral triangle.

See

8 and 14

Amazing Circle Geometry

Geometry

Triangle Classification by Angles

A right triangle is a triangle having a right angle (see Angles).

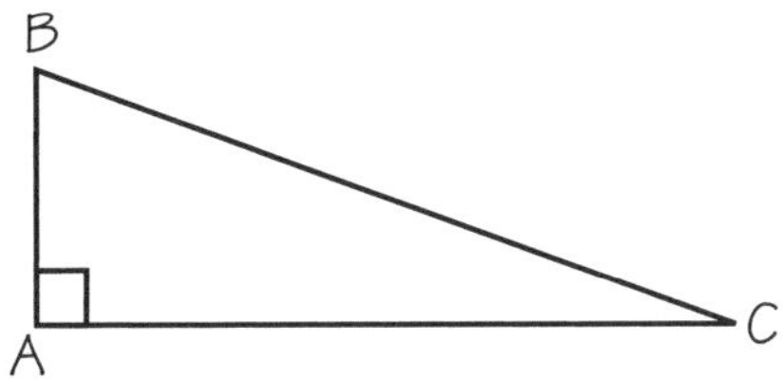

In Δ ABC, ∠ CAB is the right angle. The side opposite the right angle, $\overline{BC}$, is called the hypotenuse. The perpendicular sides are called legs.

An obtuse triangle is a triangle having an obtuse angle (see Angles).

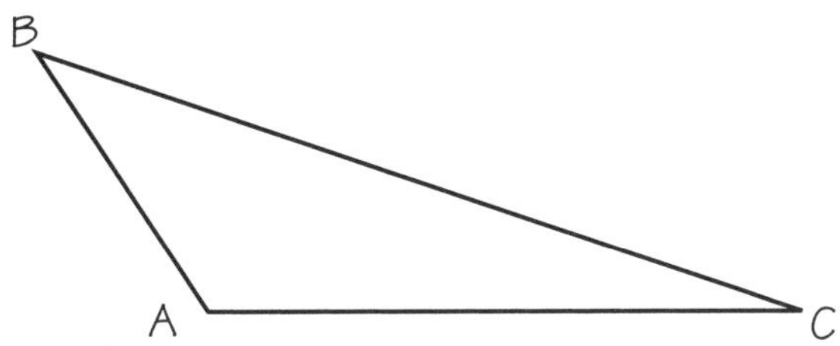

In Δ ABC, ∠ CAB is the obtuse angle.

An acute triangle is a triangle having three acute angles.

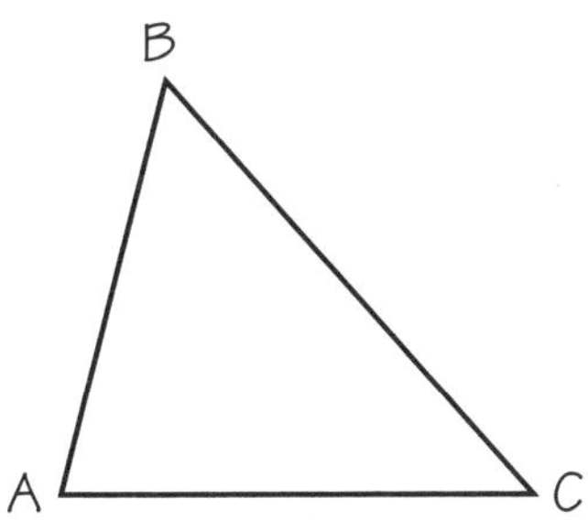

In Δ ABC, ∠CAB, ∠ABC, and ∠BCA are acute angles.

The Amazing Circle

Triangle Classification by Angles

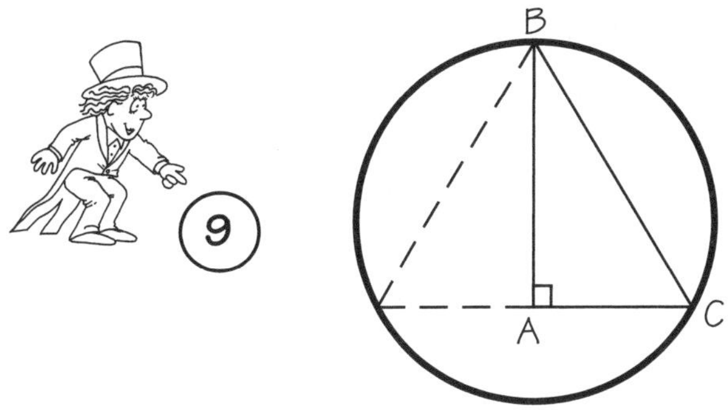

∠BAC is a right angle.
Δ ABC is a right triangle.
Δ ABC is a 30°, 60°, 90° right triangle.

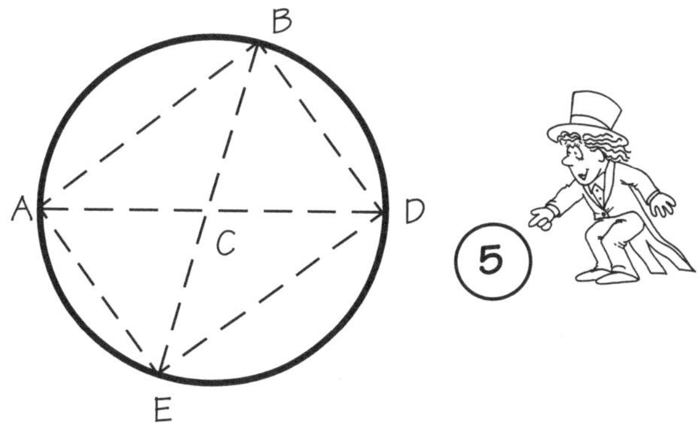

∠ACB is an obtuse angle.
Δ ABC is an obtuse triangle.

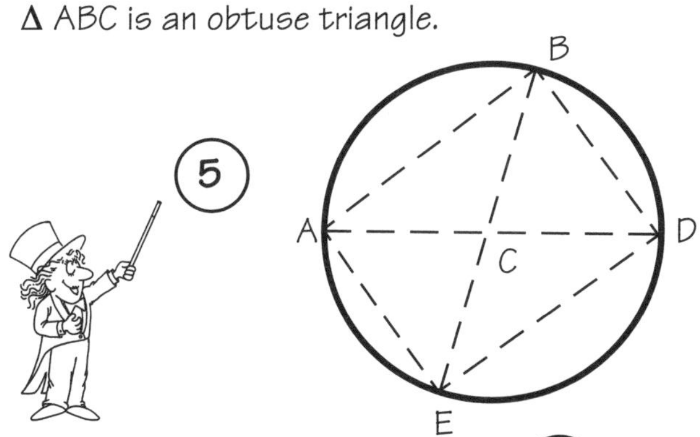

∠ACE is an acute angle.
Δ ACE is an acute triangle.

See 

and 

Amazing Circle Geometry

Geometry

Special Lines In A Triangle

Angle Bisector

The line that bisects an angle of a triangle and extends to the side opposite the angle is called an angle bisector.

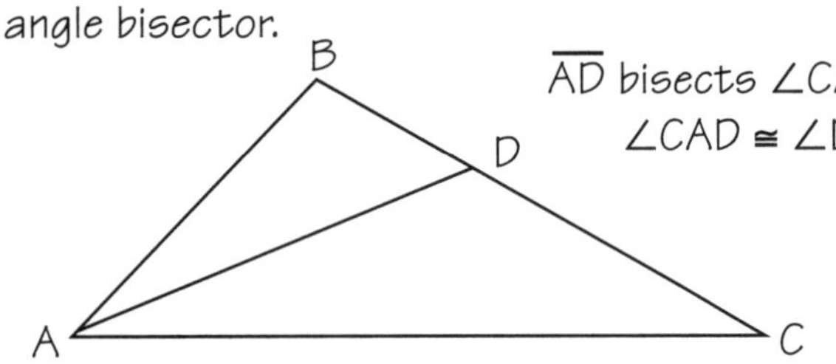

$\overline{AD}$ bisects ∠CAB.
∠CAD ≅ ∠DAB

Median

The line from the vertex of an angle to the midpoint of the side opposite the angle is called a median.

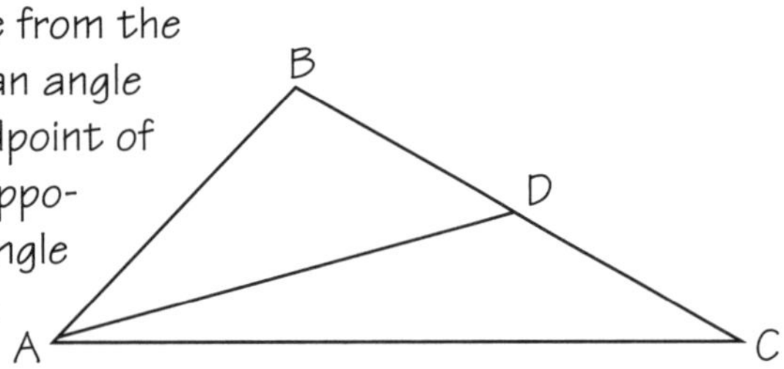

D is the midpoint of $\overline{BC}$. $\overline{AD}$ is the median.

Altitude

The line from a vertex that is perpendicular to the side opposite the vertex is called an altitude.

$\overline{BD}$ is an altitude of Δ ABC.

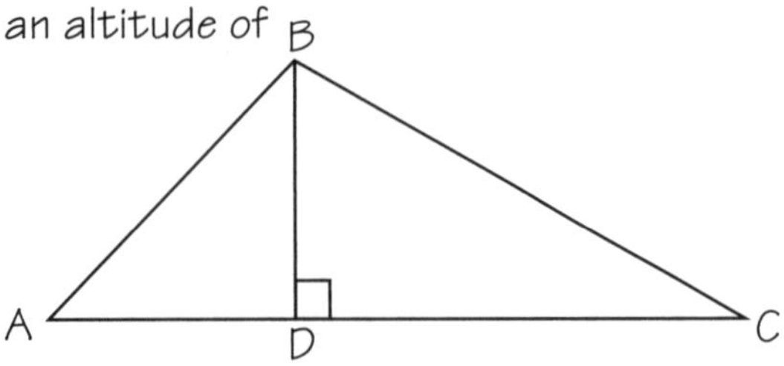

In an obtuse triangle the altitudes on either side of the obtuse angle fall outside of the triangle.

$\overline{AD}$ and $\overline{CE}$ are altitudes of Δ ABC.

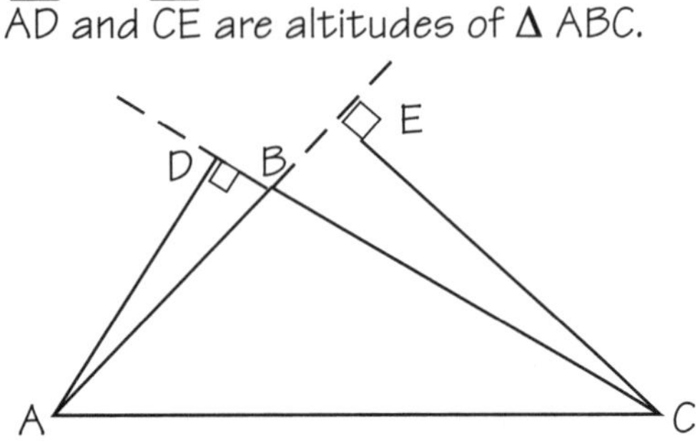

The Amazing Circle

Special Lines In A Triangle

Angle Bisector

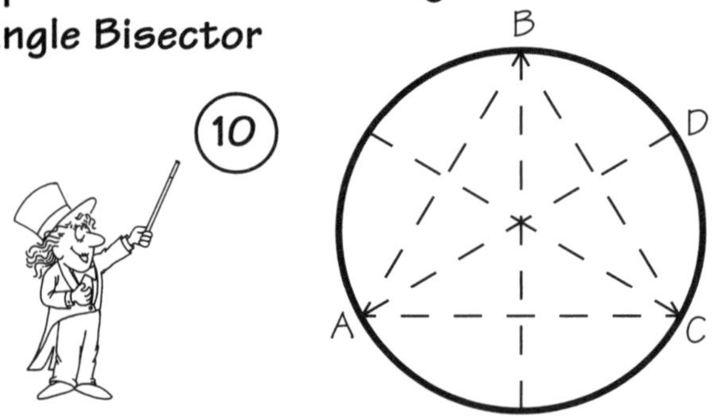

$\overline{AD}$ bisects ∠ CAB. ∠CAD ≅ ∠DAB

Median

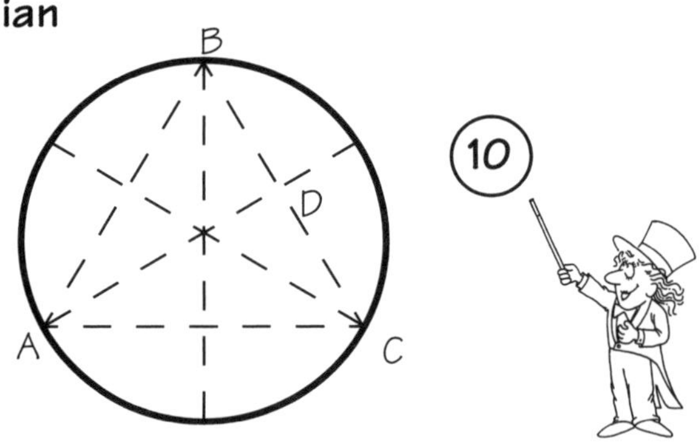

D is the midpoint of $\overline{BC}$.
$\overline{AD}$ is the median.

Altitude

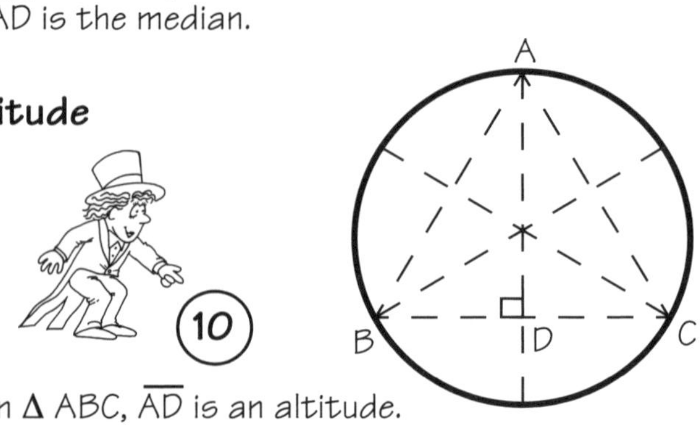

In Δ ABC, $\overline{AD}$ is an altitude.

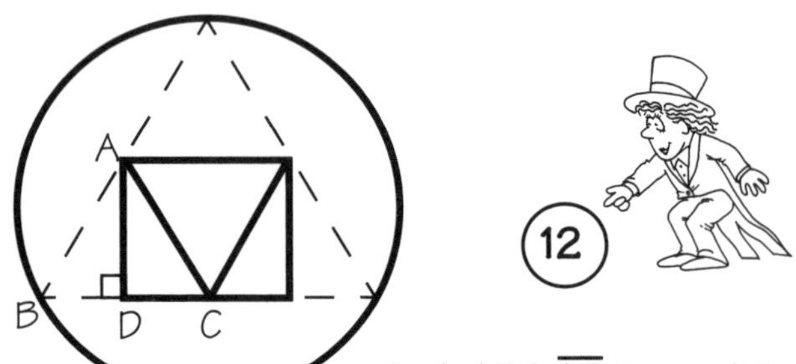

In Δ ABC $\overline{AD}$ is an altitude.

Amazing Circle Geometry

Geometry

Quadrilateral

A polygon with exactly four sides is called a quadrilateral.

Trapezoid

A quadrilateral in which one and only one pair of sides is parallel is called a trapezoid.

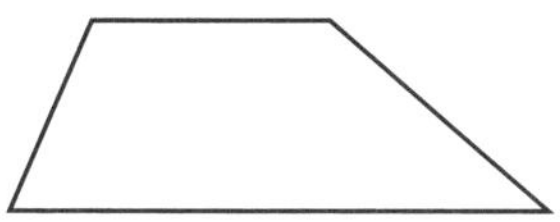

Isosceles Trapezoid

A trapezoid in which the nonparallel sides are equal in length is called an isosceles trapezoid.

$\overline{AD} \cong \overline{BC}$

$\overline{AB} \parallel \overline{DC}$

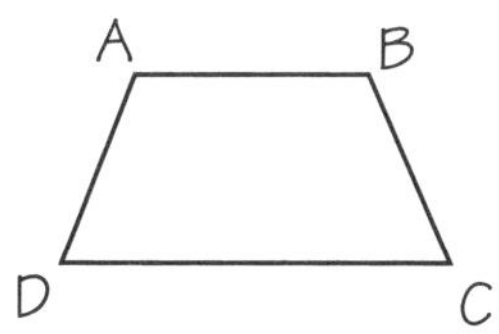

Parallelogram

A parallelogram is a quadrilateral in which both pairs of opposite sides are equal. The opposite angles are also equal.

$\overline{AC} \cong \overline{CD}$
$\overline{EC} \cong \overline{CB}$
$\overline{AB} \cong \overline{ED}$
$\overline{AE} \cong \overline{BD}$
$\angle AED \cong \angle ABD$
$\angle EAB \cong \angle EDB$

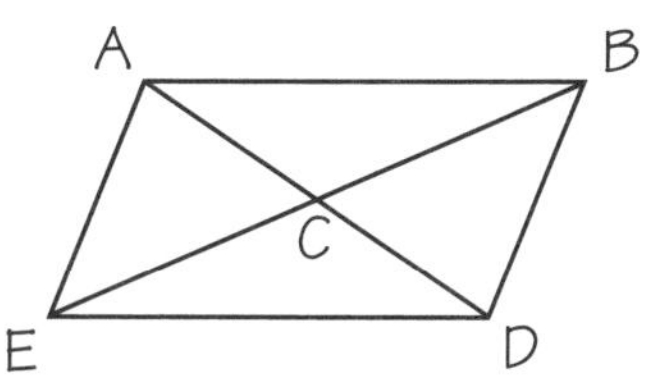

Rhombus

A parallelogram with four equal sides is called a rhombus.

$\overline{AB} \cong \overline{BD} \cong \overline{DE} \cong \overline{EA}$
$\angle EAB \cong \angle EDB$
$\angle AED \cong \angle DBA$

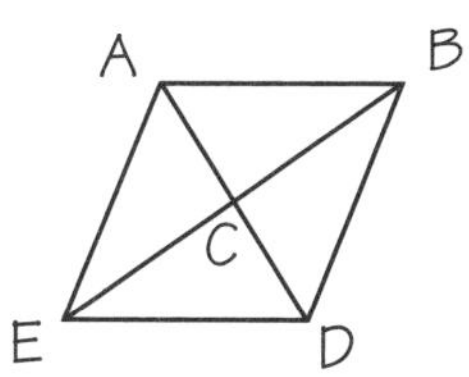

The Amazing Circle

Isosceles Trapezoid

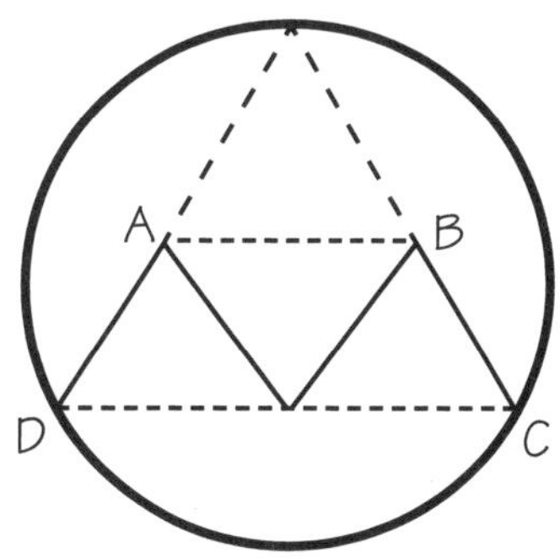

$\overline{AD} \cong \overline{BC}$

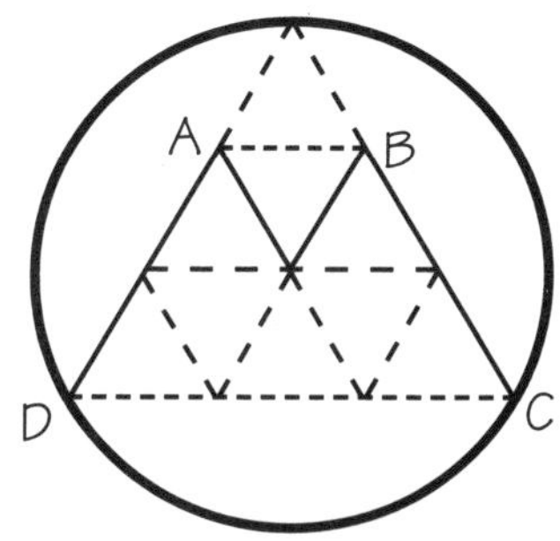

$\overline{AD} \cong \overline{BC}$

Rhombus

$\angle DAB \cong \angle BCD$
$\angle ADC \cong \angle ABC$
$\overline{AB} \cong \overline{BC} \cong \overline{CD} \cong \overline{DA} \cong \overline{AC}$

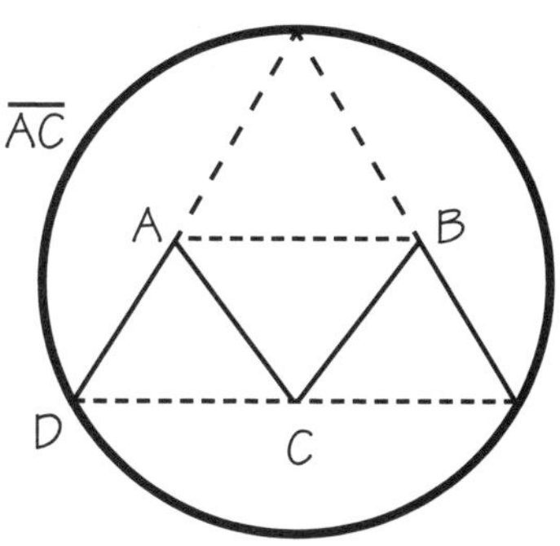

Amazing Circle Geometry

Geometry

Rectangle

A parallelogram in which one angle equals 90° is called a rectangle.

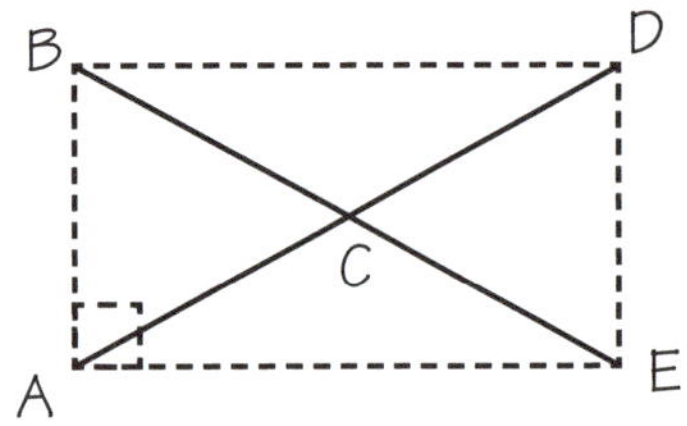

$\overline{AB} \cong \overline{ED}$

$\overline{BD} \cong \overline{AE}$

$\overline{AD} \cong \overline{BE}$

$\overline{AC} \cong \overline{CE} \cong \overline{BC} \cong \overline{CD}$

$\Delta ACB \cong \Delta ECD$

$\Delta BCD \cong \Delta ACE$

$\angle A \cong \angle B \cong \angle C \cong \angle E \cong 90°$.

Square

A rectangle in which all four sides are equal is called a square.

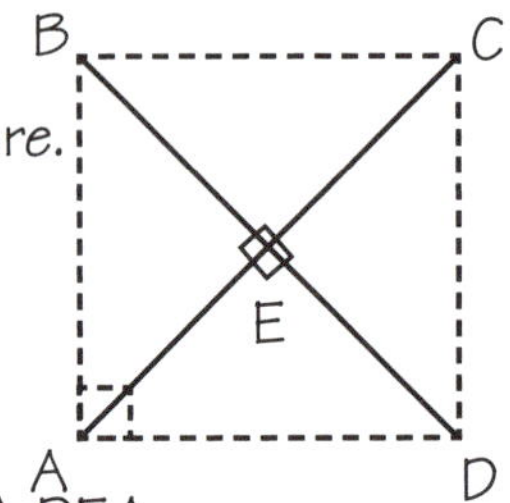

$\overline{AB} \cong \overline{BC} \cong \overline{CD} \cong \overline{DA}$

$\overline{AC} \cong \overline{BD}$

$\overline{AC} \perp \overline{BD}$

$\overline{AE} \cong \overline{EC} \cong \overline{BE} \cong \overline{ED}$

$\Delta AEB \cong \Delta BEC \cong \Delta CED \cong \Delta DEA$

$\angle A \cong \angle B \cong \angle C \cong \angle D \cong 90°$.

The diagram shows that every square is a rhombus, a rectangle, and a parallelogram.

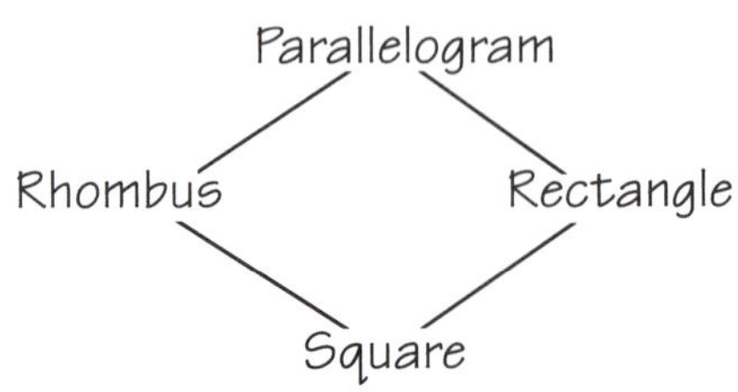

Diagonal Properties:				
1. bisect each other	✓	✓	✓	✓
2. are congruent		✓		✓
3. are perpendicular			✓	✓
4. bisect vertex angles			✓	✓
5. form 2 pairs of congruent triangles	✓	✓	✓	✓
6. form 4 congruent triangles			✓	✓

The Amazing Circle

Rectangle

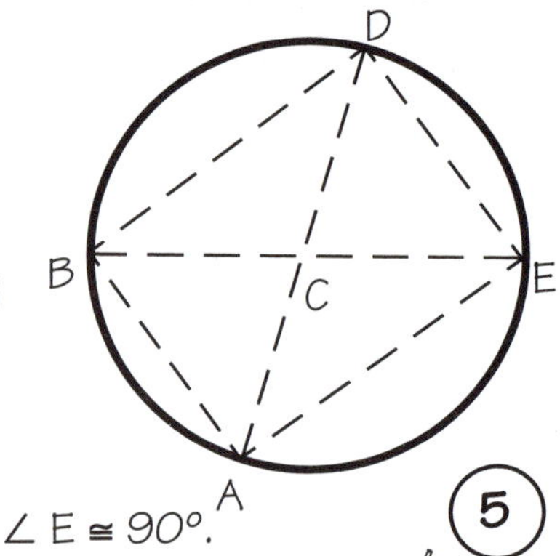

$\overline{AB} \cong \overline{ED}$

$\overline{BD} \cong \overline{AE}$

$\overline{AD} \cong \overline{BE}$

$\overline{AC} \cong \overline{CE} \cong \overline{BC} \cong \overline{CD}$

$\Delta ACB \cong \Delta ECD$

$\Delta BCD \cong \Delta ACE$

$\angle A \cong \angle B \cong \angle C \cong \angle E \cong 90°$.

Square

$\overline{AC} \cong \overline{BD}$

$\overline{AC} \perp \overline{BD}$

$\overline{AE} \cong \overline{EC} \cong \overline{BE} \cong \overline{ED}$

$\overline{AB} \cong \overline{BC} \cong \overline{CD} \cong \overline{DA}$

$\Delta AEB \cong \Delta BEC \cong \Delta CED \cong \Delta DEA$

$\angle A \cong \angle B \cong \angle C \cong \angle D \cong 90°$.

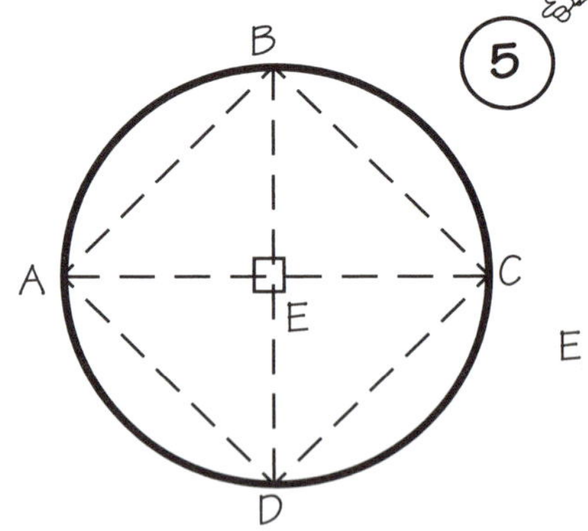

The AIMS Program

AIMS is the acronym for "**A**ctivities **I**ntegrating **M**athematics and **S**cience." Such integration enriches learning and makes it meaningful and holistic. AIMS began as a project of Fresno Pacific University to integrate the study of mathematics and science in grades K-9, but has since expanded to include language arts, social studies, and other disciplines.

AIMS is a continuing program of the non-profit AIMS Education Foundation. It had its inception in a National Science Foundation funded program whose purpose was to explore the effectiveness of integrating mathematics and science. The project directors in cooperation with 80 elementary classroom teachers devoted two years to a thorough field-testing of the results and implications of integration.

The approach met with such positive results that the decision was made to launch a program to create instructional materials incorporating this concept. Despite the fact that thoughtful educators have long recommended an integrative approach, very little appropriate material was available in 1981 when the project began. A series of writing projects have ensued, and today the AIMS Education Foundation is committed to continue the creation of new integrated activities on a permanent basis.

The AIMS program is funded through the sale of this developing series of books and proceeds from the Foundation's endowment. All net income from program and products flows into a trust fund administered by the AIMS Education Foundation. Use of these funds is restricted to support of research, development, and publication of new materials. Writers donate all their rights to the Foundation to support its on-going program. No royalties are paid to the writers.

The rationale for integration lies in the fact that science, mathematics, language arts, social studies, etc., are integrally interwoven in the real world from which it follows that they should be similarly treated in the classroom where we are preparing students to live in that world. Teachers who use the AIMS program give enthusiastic endorsement to the effectiveness of this approach.

Science encompasses the art of questioning, investigating, hypothesizing, discovering, and communicating. Mathematics is the language that provides clarity, objectivity, and understanding. The language arts provide us powerful tools of communication. Many of the major contemporary societal issues stem from advancements in science and must be studied in the context of the social sciences. Therefore, it is timely that all of us take seriously a more holistic mode of educating our students. This goal motivates all who are associated with the AIMS Program. We invite you to join us in this effort.

Meaningful integration of knowledge is a major recommendation coming from the nation's professional science and mathematics associations. The American Association for the Advancement of Science in *Science for All Americans* strongly recommends the integration of mathematics, science, and technology. The National Council of Teachers of Mathematics places strong emphasis on applications of mathematics such as are found in science investigations. AIMS is fully aligned with these recommendations.

Extensive field testing of AIMS investigations confirms these beneficial results:

1. Mathematics becomes more meaningful, hence more useful, when it is applied to situations that interest students.
2. The extent to which science is studied and understood is increased, with a significant economy of time, when mathematics and science are integrated.
3. There is improved quality of learning and retention, supporting the thesis that learning which is meaningful and relevant is more effective.
4. Motivation and involvement are increased dramatically as students investigate real-world situations and participate actively in the process.

We invite you to become part of this classroom teacher movement by using an integrated approach to learning and sharing any suggestions you may have. The AIMS Program welcomes you!

AIMS Education Foundation Programs

Practical proven strategies to improve student achievement

When you host an AIMS workshop for elementary and middle school educators, you will know your teachers are receiving effective usable training they can apply in their classrooms immediately.

Designed for teachers—AIMS Workshops:

- Correlate to your state standards;
- Address key topic areas, including math content, science content, problem solving, and process skills;
- Teach you how to use AIMS' effective hands-on approach;
- Provide practice of activity-based teaching;
- Address classroom management issues, higher-order thinking skills, and materials;
- Give you AIMS resources; and
- *Offer college (graduate-level) credits for many courses.*

Aligned to district and administrator needs—AIMS workshops offer:

- Flexible scheduling and grade span options;
- Custom (one-, two-, or three-day) workshops to meet specific schedule, topic and grade-span needs;
- Pre-packaged one-day workshops on most major topics—only $3,900 for up to 30 participants (includes all materials and expenses);
- Prepackaged *week-long* workshops (four- or five-day formats) for in-depth math and science training—only $12,300 for up to 30 participants (includes all materials and expenses);
- Sustained staff development, by scheduling workshops throughout the school year and including follow-up and assessment;
- Eligibility for funding under the Eisenhower Act and No Child Left Behind; and
- Affordable professional development—save when you schedule consecutive-day workshops.

University Credit - Correspondence Courses

AIMS offers correspondence courses through a partnership with Fresno Pacific University.

- *Convenient* distance-learning courses—you study at your own pace and schedule. No computer or Internet access required!

The tuition for each three-semester unit graduate-level course is $264 plus a materials fee.

The AIMS Instructional Leadership Program

This is an AIMS staff-development program seeking to prepare facilitators for leadership roles in science/math education in their home districts or regions. Upon successful completion of the program, trained facilitators become members of the AIMS Instructional Leadership Network, qualified to conduct AIMS workshops, teach AIMS in-service courses for college credit, and serve as AIMS consultants. Intensive training is provided in mathematics, science, process and thinking skills, workshop management, and other relevant topics.

Introducing AIMS Science Core Curriculum

Developed in alignment with your state standards, AIMS' Science Core Curriculum gives students the opportunity to build content knowledge, thinking skills, and fundamental science processes.

- *Each* grade specific module has been developed to extend the AIMS approach to full-year science programs.
- *Each* standards-based module includes math, reading, hands-on investigations, and assessments.

Like all AIMS resources these core modules are able to serve students at all stages of readiness, making these a great value across the grades served in your school.

For current information regarding the programs described above, please complete the following:

Information Request

Please send current information on the items checked:

_____ *Basic Information Packet* on AIMS materials ____ Hosting information for AIMS workshops
_____ *AIMS Instructional Leadership Program* ____ AIMS Science Core Curriculum

Name ______________________ Phone ______________________

Address ____
Street City State Zip

AIMS Magazine

YOUR K-9 MATH AND SCIENCE CLASSROOM ACTIVITIES RESOURCE

The AIMS Magazine is your source for standards-based, hands-on math and science investigations. Each issue is filled with teacher-friendly, ready-to-use activities that engage students in meaningful learning.

- *Four issues each year (October, December, February, and April).*

Current issue is shipped with all past issues within that volume.

1820	Volume XX	2005-2006	$19.95
1821	Volume XXI	2006-2007	$19.95
Two-Volume Combination			
M20507	Volumes XX & XXI	2005-2007	$34.95

Back Volumes Available
Complete volumes available for purchase:

1802	Volume II	1987-1988	$19.95
1804	Volume IV	1989-1990	$19.95
1805	Volume V	1990-1991	$19.95
1807	Volume VII	1992-1993	$19.95
1808	Volume VIII	1993-1994	$19.95
1809	Volume IX	1994-1995	$19.95
1810	Volume X	1995-1996	$19.95
1811	Volume XI	1996-1997	$19.95
1812	Volume XII	1997-1998	$19.95
1813	Volume XIII	1998-1999	$19.95
1814	Volume XIV	1999-2000	$19.95
1815	Volume XV	2000-2001	$19.95
1816	Volume XVI	2001-2002	$19.95
1817	Volume XVII	2002-2003	$19.95
1818	Volume XVIII	2003-2004	$19.95
1819	Volume XIX	2004-2005	$35.00

Call today to order back volumes: 1.888.733.2467.

Call 1.888.733.2467 or go to www.aimsedu.org

Subscribe to the AIMS Magazine

$19.95 a year!

AIMS Magazine is published four times a year.

Subscriptions ordered at any time will receive all the issues for that year.

AIMS Online – www.aimsedu.org

For the latest on AIMS publications, tips, information, and promotional offers, check out AIMS on the web at www.aimsedu.org. Explore our activities, database, discover featured activities, and get information on our college courses and workshops, too.

AIMS E-News

While visiting the AIMS website, sign up for AIMS E-News, our FREE e-mail newsletter. Published four times per year, AIMS E-News brings you food for thought. Each issue delivers:

- **Thought-provoking articles on curriculum and pedagogy;**
- **Creative ideas on teaching math and science more effectively; and**
- **Sample activities.**

Sign up today!

AIMS Program Publications

Actions with Fractions 4-9
Awesome Addition and Super Subtraction 2-3
Bats Incredible! 2-4
Brick Layers 4-9
Brick Layers II 4-9
Chemistry Matters 4-7
Counting on Coins K-2
Cycles of Knowing and Growing 1-3
Crazy about Cotton Book 3-7
Critters K-6
Down to Earth 5-9
Electrical Connections 4-9
Exploring Environments Book K-6
Fabulous Fractions 3-6
Fall into Math and Science K-1
Field Detectives 3-6
Finding Your Bearings 4-9
Floaters and Sinkers 5-9
From Head to Toe 5-9
Fun with Foods 5-9
Glide into Winter with Math & Science K-1
Gravity Rules! Activity Book 5-12
Hardhatting in a Geo-World 3-5
It's About Time K-2
It Must Be A Bird Pre-K-2
Jaw Breakers and Heart Thumpers 3-5
Problem Solving: Just for the Fun of It! 4-9
Looking at Geometry 6-9
Looking at Lines 6-9
Machine Shop 5-9
Magnificent Microworld Adventures 5-9
Marvelous Multiplication and Dazzling Division 4-5
Math + Science, A Solution 5-9
Mostly Magnets 2-8
Movie Math Mania 6-9
Multiplication the Algebra Way 4-8
Off The Wall Science 3-9
Our Wonderful World 5-9
Out of This World 4-8
Overhead and Underfoot 3-5
Paper Square Geometry:
The Mathematics of Origami
Puzzle Play: 4-8
Pieces and Patterns 5-9
Popping With Power 3-5
Primarily Bears K-6
Primarily Earth K-3
Primarily Physics K-3
Primarily Plants K-3
Proportional Reasoning 6-9
Ray's Reflections 4-8
Sense-Able Science K-1
Soap Films and Bubbles 4-9
Spatial Visualization 4-9
Spills and Ripples 5-12
Spring into Math and Science K-1
The Amazing Circle 4-9
The Budding Botanist 3-6
The Sky's the Limit 5-9
Through the Eyes of the Explorers 5-9
Under Construction K-2
Water Precious Water 2-6
Weather Sense:
Temperature, Air Pressure, and Wind 4-5
Weather Sense: Moisture 4-5
Winter Wonders K-2

Spanish/English Editions*

Brinca de alegria hacia la Primavera con las
Matemáticas y Ciencias K-1
Cáete de gusto hacia el Otoño con las
Matemáticas y Ciencias K-1
Conexiones Eléctricas 4-9
El Botanista Principiante 3-6
Los Cinco Sentidos K-1
Ositos Nada Más K-6
Patine al Invierno con Matemáticas y Ciencias K-1
Piezas y Diseños 5-9
Primariamente Física K-3
Primariamente Plantas K-3
Principalmente Imanes 2-8

* All Spanish/English Editions include student pages in Spanish and teacher and student pages in English.

Spanish Edition

Constructores II: Ingeniería Creativa Con Construcciones LEGO® (4-9)
The entire book is written in Spanish. English pages not included.

Other Science and Math Publications

Historical Connections in Mathematics, Vol. I 5-9
Historical Connections in Mathematics, Vol. II 5-9
Historical Connections in Mathematics, Vol. III 5-9
Mathematicians are People, Too
Mathematicians are People, Too, Vol. II
Teaching Science with Everyday Things
What's Next, Volume 1, 4-12
What's Next, Volume 2, 4-12
What's Next, Volume 3, 4-12

For further information write to:
AIMS Education Foundation • P.O. Box 8120 • Fresno, California 93747-8120
www.aimsedu.org/ • Fax 559•255•6396

AIMS Duplication Rights Program

AIMS has received many requests from school districts for the purchase of unlimited duplication rights to AIMS materials. In response, the AIMS Education Foundation has formulated the program outlined below. There is a built-in flexibility which, we trust, will provide for those who use AIMS materials extensively to purchase such rights for either individual activities or entire books.

It is the goal of the AIMS Education Foundation to make its materials and programs available at reasonable cost. All income from the sale of publications and duplication rights is used to support AIMS programs; hence, strict adherence to regulations governing duplication is essential. Duplication of AIMS materials beyond limits set by copyright laws and those specified below is strictly forbidden.

Limited Duplication Rights

Any purchaser of an AIMS book may make up to *200 copies* of any activity in that book for use at *one school site.* Beyond that, rights must be purchased according to the appropriate category.

Unlimited Duplication Rights for Single Activities

An individual or school may purchase the right to make an unlimited number of copies of a single activity. The royalty is $5.00 per activity per school site.

Examples: 3 activities x 1 site x $5.00 = $15.00
9 activities x 3 sites x $5.00 = $135.00

Unlimited Duplication Rights for Entire Books

A school or district may purchase the right to make an unlimited number of copies of a single, *specified* book. The royalty is $20.00 per book per school site. This is in addition to the cost of the book.

Examples: 5 books x 1 site x $20.00 = $100.00
12 books x 10 sites x $20.00 = $2400.00

Magazine/Newsletter Duplication Rights

Any purchaser of the *AIMS Magazine/Newsletter* is hereby granted permission to make up to 200 copies of any portion of it, provided these copies will be used for educational purposes.

Workshop Instructors' Duplication Rights

Workshop instructors may distribute to registered workshop participants a maximum of 100 copies of any article and/or 100 copies of no more than eight activities, provided these six conditions are met:

1. Since all AIMS activities are based upon the *AIMS Model of Mathematics* and the *AIMS Model of Learning*, leaders must include in their presentations an explanation of these two models.
2. Workshop instructors must relate the AIMS activities presented to these basic explanations of the AIMS philosophy of education.
3. The copyright notice must appear on all materials distributed.
4. Instructors must provide information enabling participants to apply for membership in the AIMS Education Foundation or order books from the Foundation.
5. Instructors must inform participants of their limited duplication rights as outlined below.
6. Only student pages may be duplicated.

Written permission must be obtained for duplication beyond the limits listed above. Additional royalty payments may be required.

Workshop Participants' Rights

Those enrolled in workshops in which AIMS student activity sheets are distributed may duplicate a maximum of 35 copies or enough to use the lessons one time with one class, whichever is less. Beyond that, rights must be purchased according to the appropriate category.

Application for Duplication Rights

The purchasing agency or individual must clearly specify the following:

1. Name, address, and telephone number
2. Titles of the books for Unlimited Duplication Rights contracts
3. Titles of activities for Unlimited Duplication Rights contracts
4. Names and addresses of school sites for which duplication rights are being purchased.

NOTE: Books to be duplicated must be purchased separately and are not included in the contract for Unlimited Duplication Rights.

The requested duplication rights are automatically authorized when proper payment is received, although a *Certificate of Duplication Rights* will be issued when the application is processed.

Address all correspondence to: **Contract Division**
AIMS Education Foundation
P.O. Box 8120
Fresno, CA 93747-8120